*Introduction to Path-Integral Methods*
*in Physics and Polymer Science*

# Introduction to Path-Integral Methods in Physics and Polymer Science

FW Wiegel

*Center for Theoretical Physics*
*Twente University of Technology*
*Enschede, The Netherlands*

World Scientific

*Published by*

World Scientific Publishing Co Pte Ltd.
P. O. Box 128, Farrer Road, Singapore 9128
242 Cherry Street, Philadelphia PA 19106-1906, USA

Library of Congress Cataloging-in-Publication Data

Weigle, F. W.
    Introduction to path-integral methods in physics
and polymer science.

    Includes bibliographies.
    1. Integrals, Path.   2. Polymers and polymerization.
I. Title.
QC174.17.P27W45    1986        530.1′5        85–29641
ISBN  9971-978-70-9

Printed in Singapore by Fu Loong Lithographer Pte Ltd.

*Dedicated to C.A. and J.K.*

# Preface

This monograph grew out of sets of lecture notes which I wrote on various occasions, always essentially for my own enjoyment. Its goal is to develop in detail those path-integral methods which have proved themselves to be of use in physics and related sciences and to demonstrate some of its most important applications. It is not a mathematical textbook filled with existence theorems, but more similar to a collection of useful recipes: a cookbook on path integration, written by an admirer who happens to be a theoretical physicist and meant for physicists, polymer scientists, applied mathematicians and others.

To date the only existing texts on path integration are the two excellent books by Feynman and Hibbs and by Schulman. The first of these is now some twenty years old and much precious material has accumulated since it appeared. The second is a rather advanced textbook, and I have felt the need to write a more elementary introduction to this subject to make it accessible to as many applied scientists as possible.

In the following pages the term *path integral* is used interchangeably with the terms *functional integral*, *integral over trajectories* or *integral over histories*. Although some authors have advocated using these different terms for different mathematical objects, a somewhat loose terminology seems justified because the various representations of a physical quantity can usually be transformed into each other.

I have made an attempt to structure the material in such a way that a smooth transition comes about from the elementary to more complicated matters. Chapters I and II form an elementary introduction to the Wiener path integral as applied to Brownian motion. In the same way Chapter III is an introduction to the applications to the statistics of polymers and membranes. These first three chapters should be intelligible to an advanced undergraduate student and assume only some basic knowledge about random walks and the diffusion equation. The next three chapters elaborate on this material by discussing further applications to polymer entanglements, the Aharonov-Bohm effect in quantum mechanics and classical statistical physics. Chapters V and VI are roughly at the level of a beginning graduate student in physics. The next three chapters (VII-IX) are an introduction to problems in quantum statistical physics and to the conceptual understanding of the behavior of systems with very many degrees of freedom. I hope Chapters IV and VII-IX will be of use to graduate students and other research workers as a starting point for the study of the more advanced textbooks mentioned before, and of the journal literature. In the final chapter, which is written in the spirit of natural philosophy, I have tried to situate path integration in the much wider perspective brought about by the ongoing development of theoretical physics towards a more holistic and organic structure.

I have presented various parts of this material in lectures which I gave at the NATO Advanced Study Institute on Path Integrals at the University of Antwerp in 1977, in lectures presented as a visiting professor at the Laboratoire de Physique Théorique of the Ecole Polytechnique Fédérale de Lausanne in 1978, in lectures presented as a visiting professor at the Department of Chemistry of Stanford University in 1979, and in seminars at the Los Alamos National Laboratory and other research institutes during the last fifteen years. It is a pleasure to thank the many students, friends and colleagues who helped me keep my love for this subject alive throughout the years; I would like to mention especially Mark Kac, Arnold Siegert, Jaap Hijmans, Michael Revzen, Jozef Devreese, Philippe Choquard, John Ross and George Bell. I am also grateful to Elly Reimerink for careful typing of the manuscript.

This book is dedicated to the two individuals who were the greatest creative forces in my private life. Although neither of them is a physicist, the reader would act in their spirit if he or she would communicate to me any thoughts or suggestions concerning the contents of or the form of presentations followed in this book.

*Amsterdam*                                                                          F.W.W.
*Fall*, 1985

# Contents

# I.  INTRODUCTION

## 1.1.  The early history of path integration

The subject of this monograph, the application of path-integral methods to problems in theoretical physics, has deep roots in the world view of contemporary physics. In a strictly mathematical sense the phrase path integration simply refers to some future generalization of the integral calculus to functionals. In a physical sense the phrase path integration refers more directly to the Holy Grail of theoretical physics: a future theory which enables us to understand the behavior of systems with very many degrees of freedom, such as classical and quantum many-particle systems or field theories.

The development of a calculus for functionals began with the work of Volterra early in this century [1]. From a practical point of view the main fruit of Volterra's work is a simple and general method for handling a problem which involves functionals. As we are going to use this recipe many times we spell it out as follows. We shall call any function of infinitely many variables, which can be labelled by a continuous index, a functional. Volterra's recipe consists of three steps: (a) Replace the functional by a function of a finite number ($N$) of variables. (b) Perform all calculations with this function. (c) Take the limit in which $N$ tends to infinity. In physical applications one expects that this recipe should lead to the correct result for the following heuristic reason. Experimental observations are in principle always somewhat coarse-grained. Hence from an experimental point of view one never knows a functional, but only some discrete approximation which corresponds to some large but finite value of $N$. As the physical theory only serves to describe these coarse-grained observations discrete approximations of functionals should suffice to formulate the theory; this, however, is exactly what Volterra's recipe does.

The first attempts to integrate a functional over a space of functions can be found in some early papers by Daniell [2–5]. The historical reasons for Daniell's failure have been discussed by Kac [6]. A few years later Wiener introduced a measure in function space which is kosher from a mathematical point of view and which made it possible to define an integral of a functional over a space of functions [7–11]. This integral, which is now called the Wiener integral, has played a central role in the further development of the subject of path integration. We shall discuss various aspects of the Wiener integral in further sections of Chapter I. The reader who wants to pursue the early history of the Wiener integral should consult the paper by Kac [6] mentioned before, as well as a paper by Papadopoulos [12]. The work of Wiener did not find applications in theoretical physics, apart from some mysterious comments in a paper by Kirkwood [14] about the possible use of Wiener-like path integrals in quantum statistical mechanics.

A second development, which was going to have a great impact on theoretical physics, was initiated by a paper by Dirac [15] on the role of the Lagrangian in quantum mechanics. This eventually led Feynman [17, 18] to represent the propagator of the Schrödinger equation by the complex-valued path integral that now bears his name. The Feynman path integral is the subject of Chapter V.

The first applications of path integration to specific problems in theoretical physics were studies of Brownian motion in an absorbing medium [19] and developments in the theory of superfluid $^4$He [20–28]. This early work is the subject of the review papers of Gelfand and Yaglom [29] and Brush [30]. Other applications of path integration were to quantum field theory [31, 32] and to the theory of homogeneous turbulence [33]. After about 1955 the number of papers in which path integration played an essential role has been growing steadily; at the time of this writing I estimate their rate of creation to be about one per week. We shall, therefore, not pursue the history of the subject past the mid-fifties.

Before we turn to the study of the Wiener integral I would like to make some remarks which are meant to help the reader to cope with the literature. The classical monograph on path integration is the well-known book by Feynman and Hibbs [34]. A more recent text is the book by Schulman [35]. Both works are well written and any serious student of the subject is advised to read them. Some of the original research papers to which we refer in these pages are well written too—especially those by Feynman himself—but of many others all one can say is that their reading is so unpleasant that it borders on masochism. On such occasions the reader might prefer to first browse through the review papers about this topic, which are much more accessible, and only after that turn to the original papers. Some of the main review papers are those by Feynman [18], Kac [19], Gelfand and Yaglom [29], Brush [30], Garrod [36], Wiegel [37], Neveu [38], Dashen [39] and DeWitt-Morette, Maheshwari and Nelson [40]. A reader in search of a research topic might also like to consult the proceedings of three conferences exclusively devoted to this subject; they are edited by Arthurs [41], Papadopoulos and Devreese [13] and by Albeverio, Combe and Hoegh-Krohn [42].

## 1.2.  Free Brownian motion and the Wiener integral

Consider a large number of particles which perform Brownian motion along some axis and which do not interact with each other. Let $c(x, t)\,dx$ denote the number of particles in a small interval $dx$ around position $x$, at time $t$. It is an experimental fact that the particle current $j(x, t)$—defined as the net number of Brownian particles that pass the point $x$ in the direction of increasing values of $x$ per unit of time—is proportional to the gradient of their concentration

$$j(x, t) = -D \frac{\partial c}{\partial x} . \tag{2.1}$$

This relation also serves as the definition of the diffusion coefficient $D$. If particles are neither created nor destroyed the concentration obeys the continuity equation

$$\frac{\partial c}{\partial t} = -\frac{\partial j}{\partial x} , \tag{2.2}$$

which can also be written in the form

$$\frac{\partial c}{\partial t} = D \frac{\partial^2 c}{\partial x^2} . \tag{2.3}$$

If we start at $t = t_0$ with one particle at $x = x_0$ the solution is denoted by $G_0(x, t)$. With the initial condition

$$G_0(x, t_0) = \delta(x - x_0) \qquad (2.4)$$

the solution of the diffusion equation is

$$G_0(x, t) = \left\{ 4\pi D(t - t_0) \right\}^{-1/2} \exp \left\{ -\frac{(x - x_0)^2}{4D(t - t_0)} \right\}, \qquad (t \geq t_0) , \qquad (2.5)$$

as can be verified by substitution. The symbol $\delta(x)$ denotes Dirac's delta function.

This has all been known since the beginning of the century and forms the subject of several textbooks on Brownian motion (cf. Einstein [43], Chandrasekhar [44], Wax [45], Barber and Ninham [46]). The only physics that goes into Brownian motion theory is Eq. (2.1). The solution (2.5) would be called the Green function of the diffusion equation by mathematicians, and the propagator of a Brownian particle by physicists.

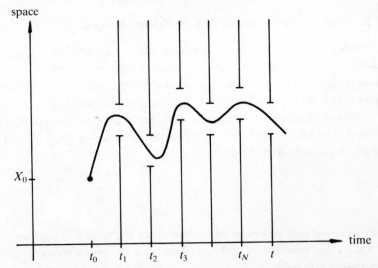

Fig. 1.1. Particle trajectory which starts at position $x_0$ at time $t_0$ and passes through $N+1$ gates $dx_1, dx_2, \ldots, dx_N, dx$ at times $t_1, t_2, \ldots, t_N, t$. The times are equally spaced.

Now suppose one divides the time interval $(t_0, t)$ into $N + 1$ equal intervals of length $\varepsilon$, separated by time points $t_1 < t_2 < \ldots < t_N$ and asks for the probability to find a particle (which started in $x_0$ at $t_0$) at coordinates $x_1, x_2, \ldots x_N, x$ at times $t_1, t_2, \ldots t_N, t$ with uncertainties $dx_1, dx_2, \ldots dx_N, dx$. This probability would obviously equal a product of propagators

(2.5) over the successive subintervals (see Fig. 1.1)

$$(4\pi D\varepsilon)^{-(N+1)/2} \exp\left\{-\frac{1}{4D\varepsilon}\sum_{j=0}^{N}(x_{j+1}-x_j)^2\right\}\prod_{j=1}^{N+1}dx_j\,, \qquad (2.6)$$

where $x_{N+1}$ is defined to equal $x$. In the limit $\varepsilon \to 0, N \to \infty$, $(N+1)\varepsilon = t - t_0$ this expression can be interpreted as the probability for the particle to follow the particular path $x(\tau)$ from $x_0$ to $x$, which is specified by $x(t_j) = x_j$. In that limit the exponential can be written in the elegant—but somewhat misleading—continuous notation

$$\exp\left\{-\frac{1}{4D}\int_{t_0}^{t}\left(\frac{dx}{d\tau}\right)^2 d\tau\right\}. \qquad (2.7)$$

Of course, by integrating the expression (2.6) over all the intermediate coordinates $x_1, x_2,\ldots x_N$ one should recover the original probability density (2.5). In this way we have derived a very basic formula which expresses a physical quantity (the propagator of a Brownian particle) as a path integral (the Wiener integral)

$$\int_{x_0,t_0}^{x,t}\exp\left\{-\frac{1}{4D}\int_{t_0}^{t}\left(\frac{dx}{d\tau}\right)^2 d\tau\right\}d[x(\tau)] = \{4\pi D(t-t_0)\}^{-1/2}$$

$$\times \exp\left\{-\frac{(x-x_0)^2}{4D(t-t_0)}\right\}. \qquad (2.8)$$

It should be emphasized that the left-hand side of this equation is merely a symbol for Volterra's recipe applied to this simple case:

(a)   The symbol $d[x(\tau)]$ indicates the "infinitesimally small" collection of functions $x(\tau)$ which obey the relations

$$\left.\begin{array}{l}x(t_0) = x_0\,, \\ \quad x_1 < x(t_1) < x_1 + dx_1\,, \\ \quad x_2 < x(t_2) < x_2 + dx_2\,, \\ \quad \vdots \qquad \vdots \qquad \vdots \\ \quad x_N < x(t_N) < x_N + dx_N\,, \\ x(t) = x\,. \end{array}\right\} \qquad (2.9)$$

This can also be expressed by

$$d[x(\tau)] <=> \prod_{j=1}^{N}dx_j\,. \qquad (2.10)$$

(b)   The path-integral symbol indicates an integration over all $dx_j$ from $-\infty$ to $+\infty$ and multiplication with the normalization factor

$$\int d[x(\tau)] <=> (4\pi\varepsilon D)^{-(N+1)/2} \int_{-\infty}^{+\infty} dx_1 \int_{-\infty}^{+\infty} dx_2 \dots \int_{-\infty}^{+\infty} dx_N \ . \qquad (2.11)$$

(c)   The fact that all functions in the integrand assume the same values at the end points is indicated in (2.8) by the "boundaries" $x_0$, $t_0$ and $x$, $t$ on the integral sign.

(d)   The continuous integrand in (2.8) is shorthand for

$$\int_{t_0}^{t} \left(\frac{dx}{d\tau}\right)^2 d\tau <=> \frac{1}{\varepsilon} \sum_{j=0}^{N} (x_{j+1} - x_j)^2 \ . \qquad (2.12)$$

(e)   After all integrations have been performed one takes the limit $N \to \infty$.

For the record only it should be noted here that there are several minor differences in the notation which physicists use for path integrals. For example, our expression

$$\int_{x_0,t_0}^{x,t} \exp\left\{-\frac{1}{4D}\int_{t_0}^{t}\left(\frac{dx}{d\tau}\right)^2 d\tau\right\} d[x(\tau)] \qquad (2.13a)$$

would be written by Feynman and Hibbs [34] as

$$\int_{a}^{b} \exp\left\{-\frac{1}{4D}\int\left(\frac{dx}{d\tau}\right)^2 d\tau\right\} \mathscr{D}x(\tau) \ , \qquad (2.13b)$$

where obviously $a \equiv x_0, t_0$ and $b \equiv x, t$. Schulman [35] would often write for the same expression

$$C \sum_{x(\cdot)} \exp\left\{-\frac{1}{4D}\int\left(\frac{dx(\cdot)}{d\tau}\right)^2 d\tau\right\} \ , \qquad (2.13c)$$

where $C$ denotes the normalization factor $(4\pi\varepsilon D)^{-(N+1)/2}$ which we have absorbed into the path-integral symbol. In practice these notational differences seldom lead to confusion.

## 1.3.   Cell representation and spectral representation

One of the most important peculiarities of path integrals is the propensity of our intuition to conceive them as sums over paths that are somehow too "smooth," whereas they are in reality sums over highly erratic objects. For example, Eq. (2.6) shows that the discretized paths $(x_0, x_1, \ldots, x_N, x)$ which contribute appreciably to (2.8) have $|x_{j+1} - x_j| = O(D^{1/2}\varepsilon^{1/2})$. For these paths $dx/d\tau$ will be of the order of magnitude $D^{1/2}\varepsilon^{-1/2}$, which tends to $\pm\infty$ in the limit $\varepsilon \to 0$. The continuous notation (2.7, 8) is therefore misleading in the sense that all the important paths are non-differentiable in the continuous limit. This well-known property of the trajectories of Brownian particles—sometimes called their fractal nature [52, 53]—has important consequences for the physics of chemoreception, for example, and it is no exaggeration to say that chemoreception by living cells would be impossible without it (cf. recent papers by Berg and Purcell [47], DeLisi and Wiegel [48] and Wiegel [50, 51]). In this way Nature has taken advantage of a phenomenon that causes the student of path integration much trouble.

A simple example is our previous result (2.8) for $x_0 = x = t_0 = 0$: let

$$G_0(t) \equiv \lim (4\pi D\varepsilon)^{-(N+1)/2} \int_{-\infty}^{+\infty} dx_1 \int_{-\infty}^{+\infty} dx_2 \ldots \int_{-\infty}^{+\infty} dx_N$$

$$\times \exp\left\{ -\frac{1}{4D\varepsilon} \sum_{j=0}^{N} (x_{j+1} - x_j)^2 \right\} . \tag{3.1}$$

Suppose we do not yet know that this Wiener integral has the value $(4\pi Dt)^{-1/2}$. Then, in order to calculate it one can follow two methods.

The first method would use what we shall call the spectral representation, in which one writes the exponential as a bilinear form

$$\sum_{j=0}^{N} (x_{j+1} - x_j)^2 = \sum_{k,l=1}^{N} x_k A_{k,l} x_l , \tag{3.2}$$

where

$$A = \begin{bmatrix} 2 & -1 & & & & \\ -1 & 2 & -1 & & & \\ & -1 & 2 & -1 & & \\ & & & & 2 & -1 \\ & & & & -1 & 2 \end{bmatrix} . \tag{3.3}$$

The $N \times N$ matrix $A_{k,l}$ has zero matrix elements apart from those in the main diagonal and in the two neighboring diagonals. As $A$ is hermitian its eigenvalues $\lambda_j$ are real and the transformation of the integral (3.1) from the coordinates $x_1, x_2, \ldots, x_N$ to the eigenvectors of $A$ should have a Jacobian equal to unity. If the expansion coefficients are denoted by $y_j$ one finds in this way

$$G_0(t) = \lim (4\pi D\varepsilon)^{-(N+1)/2} \int_{-\infty}^{+\infty} dy_1 \int_{-\infty}^{+\infty} dy_2 \ldots \int_{-\infty}^{+\infty} dy_N$$

$$\times \exp\left\{-\frac{1}{4D\varepsilon} \sum_{j=1}^{N} \lambda_j y_j^2\right\}$$

$$= \lim (4\pi D\varepsilon)^{-(N+1)/2} \prod_{j=1}^{N} (4\pi D\varepsilon/\lambda_j)^{1/2}$$

$$= \lim (4\pi D\varepsilon \, \det A_N)^{-1/2} , \tag{3.4}$$

where the subscript $N$ indicates that $A$ is an $N \times N$ matrix. All that is needed now is the determinant of the matrix $A_N$, which can be found as follows. First one calculates det $A_N$ "by hand" for small $N$. One finds det $A_1 = 2$, det $A_2 = 3$. This leads one to guess that

$$\det A_N = N + 1. \tag{3.5}$$

If det $A_N$ is expanded in the elements of the last column one finds det $A_N = 2 \det A_{N-1} - A_{N-2}$. This recursion relation is satisfied by (3.5) which proves the general validity of (3.5). Combination of the last two equations gives $G_0(t) = (4\pi Dt)^{-1/2}$ in agreement with (2.8).

The second method to calculate (3.1), which we shall call the method of the cell representation, essentially amounts to doing the integrations one by one. Use the formula

$$\int_{-\infty}^{+\infty} \exp\left\{-a(x - x')^2 - b(x - x'')^2\right\} dx = \left(\frac{\pi}{a + b}\right)^{1/2}$$

$$\times \exp\left\{-\frac{ab}{a + b}(x' - x'')^2\right\} \tag{3.6}$$

for the case $x' = x_0$, $x'' = x_2$, $x = x_1$. With the two factors $(4\pi D\varepsilon)^{-1/2}$ taken into account, the integration over $x_1$ in (3.1) leads to the expression

$$(8\pi D\varepsilon)^{-1/2} \exp\left\{-\frac{(x_0 - x_2)^2}{8D\varepsilon}\right\} , \tag{3.7}$$

which is of the same form as the factors in (3.1) but with $\varepsilon$ replaced by $2\varepsilon$ throughout. Similar factors arise from the integrations over $x_3, x_5, x_7, \ldots$ One repeats this process till all integrations have been performed, which leads once more to $G_0(t) = \{4\pi D(N + 1)\varepsilon\}^{-1/2} = (4\pi Dt)^{-1/2}$. The aficionado of the renormalization group will recognize what we just did as a primitive form of renormalization, which happens to be exact for this simple example. Chapter IX is devoted to the renormalization method.

There is a sloppy way to apply the spectral method, which relies too much on the "smooth" image of the paths in our intuition, and which easily leads to erroneous results. This can be demonstrated using (3.1) as an example. First we observe that the exact, unnormalized $l$-th eigenvector of the matrix $A_N$ has the $j$-th component $\sin(\pi l j / N + 1)$, where $l, j = 1, 2, 3, \ldots, N$. The corresponding eigenvalue equals

$$\lambda_l = 2 - 2 \cos \frac{\pi l}{N + 1} \ . \tag{3.8}$$

Hence, for $l \ll N$, that is, if $\varepsilon$ is very small compared to the period of oscillation of the $l$-th eigenvector, one can approximate

$$\lambda_l \cong \left( \frac{\pi l}{N + 1} \right)^2 \ , \qquad (l \ll N) \ . \tag{3.9}$$

It is now tempting to write

$$\det A_N = \prod_{l=1}^{N} \left( 2 - 2 \cos \frac{\pi l}{N + 1} \right) \cong \prod_{l=1}^{N} \left( \frac{\pi l}{N + 1} \right)^2 \ , \quad \text{(wrong!)} \tag{3.10}$$

but this is wrong because the product extends also over values of $l$ which are of order $N$, where the approximation (3.9) does not hold. The temptation to make this type of approximation is reinforced by taking the continuous notation (2.12) too seriously and performing a partial integration. As $x(0) = x(t) = 0$ this gives

$$\int_0^t \left( \frac{dx}{d\tau} \right)^2 d\tau = - \int_0^t x \frac{d^2x}{d\tau^2} d\tau. \tag{3.11}$$

Next one solves the eigenvalue problem

$$\frac{d^2x_l}{d\tau^2} = -\beta_l x_l \ , \qquad x_l(0) = x_l(t) = 0. \tag{3.12}$$

The solution is

$$x_l(\tau) = (2/t)^{1/2} \sin \pi l \frac{\tau}{t} \ , \tag{3.13}$$

$$\beta_l = \left( \frac{\pi l}{t} \right)^2 \ , \qquad (l = 1, 2, 3, \ldots) \ . \tag{3.14}$$

Writing $\tau = j\varepsilon$, $t = (N+1)\varepsilon$ one has found the correct eigenvector, but instead of the correct eigenvalue (3.8) one has the approximation (3.9) which is not sufficient to calculate the determinant of $A_N$ to the desirable accuracy (there is an extra factor $\varepsilon^{-2}$ due to the translation of discrete quantities into continuous ones). What is wrong with the approach which leads to (3.9–14) is that the continuous point of view has been used from the beginning; rather one should find the continuous point of view as the limit of a discrete formalism.

### 1.4.  Wiener integrals, differential equations and integral equations

An interesting example of a Wiener integral arises if one considers Brownian motion in a medium in which the Brownian particle can be annihilated with a probability $A(x, t)$ per unit of time. The particle current $j$ is still given by (2.1) but instead of the continuity equation (2.2) the balance equation for the number of Brownian particles becomes

$$\frac{\partial c}{\partial t} = -\frac{\partial j}{\partial x} - Ac \ , \tag{4.1}$$

Hence the concentration of particles is governed by the partial differential equation

$$\frac{\partial c}{\partial t} = D\frac{\partial^2 c}{\partial x^2} - Ac \ , \tag{4.2}$$

and the propagator $G_A(x, t)$ is defined for $t > t_0$ as the solution of this equation under the initial condition

$$G_A (x, t_0) = \delta(x - x_0) \ . \tag{4.3}$$

This propagator can be represented by a Wiener integral which is found in the following way. Consider an arbitrary particle path $x(\tau)$ with $x(0) = x_0$, $x(t) = x$. The probability that the Brownian particle will survive this path without being absorbed equals

$$P[x(\tau)] = \exp\{-\int_{t_0}^{t} A(x(\tau), \tau)d\tau\} \ . \tag{4.4}$$

The propagator is the sum of this expression over all paths from $x_0(t_0)$ to $x(t)$:

$$G_A(x, t) = \int_{x_0, t_0}^{x, t} \exp\left\{-\frac{1}{4D}\int_{t_0}^{t}\left(\frac{dx}{d\tau}\right)^2 d\tau - \int_{t_0}^{t} A(x, \tau) \, d\tau\right\} d[x(\tau)] \ ,$$

$$(t > t_0) \tag{4.5}$$

The path integral is defined by the steps a – e of Section 1.2, with the discrete approximation

$$\exp\left\{-\int_{t_0}^{t} A(x, \tau) \, d\tau\right\} <=> \exp\left\{-\varepsilon \sum_{j=1}^{N} A(x_j, \tau_j)\right\} . \tag{4.6}$$

In the following sections path integrals of the form (4.5) will be used frequently.

In order to further elucidate the relation which exists between Wiener path integrals of the form (4.5), differential equations of the form (4.2) and integral equations we shall use the more detailed notation $G_A(x, t \mid x_0, t_0)$ to denote the propagator at all times, i.e., we define

$$G_A(x, t \mid x_0, t_0) = G_A(x, t), \qquad (t > t_0) , \tag{4.7a}$$

$$G_A(x, t \mid x_0, t_0) = 0, \qquad (t < t_0) . \tag{4.7b}$$

Because of the limiting form (4.3) of $G_A(x, t)$ for $t \downarrow t_0$ the propagator can be solved from the inhomogeneous differential equation

$$\left[\frac{\partial}{\partial t} - D \frac{\partial^2}{\partial x^2} + A(x, t)\right] G_A(x, t \mid x_0, t_0) = \delta(x - x_0) \, \delta(t - t_0) , \tag{4.8}$$

where now $t$ can have any value, and where $G_A = 0$ for $t < t_0$. Of course, a similar equation holds for the propagator of a Brownian particle which moves in a medium without absorption

$$\left[\frac{\partial}{\partial t} - D \frac{\partial^2}{\partial x^2}\right] G_0(x, t \mid x_0, t_0) = \delta(x - x_0) \, \delta(t - t_0) . \tag{4.9}$$

But this implies that $G_A$ is the solution of the integral equation

$$G_A(x, t \mid x_0, t_0) = G_0(x, t \mid x_0, t_0) - \int_{-\infty}^{+\infty} \int_{-\infty}^{+\infty} G_0(x, t \mid x', t') A(x', t')$$

$$\times \ G_A(x', t' \mid x_0, t_0) \, dx' \, dt' , \tag{4.10}$$

as can be verified by applying the differential operator $\partial/\partial t - D \, \partial^2/\partial x^2$ to both sides of the equality and using (4.9) to perform the integrations over $dx'$ and $dt'$. It is this connection between Wiener integrals, differential equations and integral equations which can be used frequently to show that physical systems that bear no resemblance to each other are nevertheless mathematically highly similar.

## 1.5. Another derivation of the differential equation for the Wiener integral

The connection between Wiener integrals, differential equations and integral equations, which formed the subject of the previous section, is so important that we want to derive it in another way. Consider the quantity

$$G_A(x, t \mid x_0, t_0) \equiv \lim (4\pi D\varepsilon)^{-(N+1)/2} \int_{-\infty}^{+\infty} dx_1 \int_{-\infty}^{+\infty} dx_2 \ldots \int_{-\infty}^{+\infty} dx_N$$

$$\times \exp\left\{-\frac{1}{4D\varepsilon} \sum_{j=0}^{N} (x_{j+1} - x_j)^2 - \varepsilon \sum_{j=1}^{N} A(x_j, t_j)\right\}, \quad (5.1)$$

which is the definition of Eq. (4.5) for $t > t_0$. For $t < t_0$ we define $G_A(x, t \mid x_0, t_0) = 0$. The limit is of course $N \to \infty$, $\varepsilon \to 0$ and $(N + 1)\varepsilon = t - t_0$. Note that we have calculated $G_0(x, t \mid x_0, t_0)$ explicitly in Section 1.3. The second exponential factor can be expanded in a power series

$$\exp\left\{-\varepsilon \sum_{j=1}^{N} A(x_j, t_j)\right\} = 1 - \varepsilon \sum_{j=1}^{N} A(x_j, t_j) + \frac{1}{2} \varepsilon^2 \sum_{j=1}^{N} \sum_{k=1}^{N}$$

$$A(x_j, t_j) A(x_k, t_k) - \cdots \quad (5.2)$$

When the integrations over the intermediate $x$-coordinates are performed first, one finds the perturbation expansion

$$G_A(x, t \mid x_0, t_0) = G_0(x, t \mid x_0, t_0) - \varepsilon \sum_{j=1}^{N} \int_{-\infty}^{+\infty} dx_j \, G_0(x, t \mid x_j, t_j)$$

$$\times A(x_j, t_j) \, G_0(x_j, t_j \mid x_0, t_0)$$

$$+ \frac{1}{2!} \varepsilon^2 \sum_{j=1}^{N} \sum_{k=1}^{N} \int_{-\infty}^{+\infty} dx_j \int_{-\infty}^{+\infty} dx_k \, G_0(x, t \mid x_j, t_j)$$

$$\times A(x_j, t_j) \, G_0(x_j, t_j \mid x_k, t_k) A(x_k, t_k) \, G_0(x_k, t_k \mid x_0, t_0) - \cdots \quad (5.3)$$

We now note that $\varepsilon \sum_j$ can be replaced by $\int_{t_0}^{t} dt_j$ in the limit $\varepsilon \to 0$, that the factor $1/N!$ can be omitted if the time variables are ordered: $t_0 < t_j < t$, $t_0 < t_k < t_j < t$, etc., and that this chronological ordering of the time variables is automatically realized by the definition $G_0(x, t \mid x_0, t_0) = 0$ if $t < t_0$. Hence the bulky expansion (5.3) simplifies to read

$$G_A(x, t \mid x_0, t_0) = G_0(x, t \mid x_0, t_0) - \int dx' \int dt' \, G_0(x, t \mid x', t')$$
$$\times A(x', t') \, G_0(x', t' \mid x_0, t_0)$$

$$+ \int dx' \int dt' \int dx'' \int dt'' \, G_0(x, t \mid x', t') \, A(x', t') \, G_0(x', t' \mid x'', t'')$$

$$\times A(x'', t'') \, G_0(x'', t'' \mid x_0, t_0) - \ldots, \qquad (5.4)$$

where all integrations run from $-\infty$ to $+\infty$. But this is just the solution of the integral equation

$$G_A(x, t \mid x_0, t_0) = G_0(x, t \mid x_0, t_0) - \int dx' \int dt' \, G_0(x, t \mid x', t') \, A(x', t')$$

$$\times G_A(x', t' \mid x_0, t_0), \qquad (5.5)$$

which is identical to (4.10). From (5.5) one recovers the differential equation (4.8) by multiplying both sides with the operator $\partial/\partial t - D\partial^2/\partial x^2$ and using (4.9).

This derivation of the differential equation from the Wiener integral by way of the integral equation has been made mathematically rigorous by Kac [19]. A formal proof that the Wiener path integral is a Lebesgue integral can be found in Wiener's paper on generalized harmonic analysis [11]. In Section 5.2 we shall again present another derivation of the relation between path integrals and differential equations.

## 1.6. The harmonic potential

Some of the most important examples of Wiener integrals of the form (4.5) arise when the exponential is an expression which is quadratic in $x$ and $\dot{x}$. The basic case is

$$A(x, \tau) = \alpha x^2, \qquad (6.1)$$

with the constant $\alpha$ positive. These path integrals appear in problems in which a particle is subject to an harmonic force. It should be clear from the discussion of Section 1.4 that they can be calculated either by solving a partial differential equation, or an integral equation. In this section we follow another method which only "works" for quadratic exponentials, but which has the advantage of great simplicity.

The steps in the calculation are as follows:

(a) With the special choice (6.1) for the annihilation probability per unit of time the exponential in Eq. (4.5) equals the negative of

$$L \equiv \int_{t_0}^{t} \left[ \frac{1}{4D} \left( \frac{dx}{d\tau} \right)^2 + \alpha x^2 \right] d\tau . \tag{6.2}$$

Calling the quantity between the angular brackets $L(x, \dot{x})$ one sees that the integral is as small as possible (for all paths passing through $x_0$ at time $t_0$ and through $x$ at time $t$) if the Euler-Lagrange equations hold, i.e., if

$$\frac{\partial L}{\partial x} - \frac{d}{d\tau} \frac{\partial L}{\partial \dot{x}} = 0 . \tag{6.3}$$

In our case this reads

$$\frac{d^2 x}{d\tau^2} = 4\alpha D x, \qquad x(t_0) = x_0, \quad x(t) = x . \tag{6.4}$$

The solution will be denoted by $x_c(\tau)$ and is found to be given by

$$x_c(\tau) = x_0 \cosh \sqrt{4\alpha D} \, (\tau - t_0) + \frac{x - x_0 \cosh \sqrt{4\alpha D} \, (t - t_0)}{\sinh \sqrt{4\alpha D} \, (t - t_0)}$$

$$\times \sinh \sqrt{4\alpha D} \, (\tau - t_0) . \tag{6.5}$$

(b) Now we try to evaluate the Wiener integral in the cell representation of Section 1.3, but instead of using $x(\tau)$ we write

$$x(\tau) = x_c(\tau) + y(\tau) . \tag{6.6}$$

Substitution of (6.5) into (6.2) gives, after one partial integration and a somewhat messy but straightforward calculation,

$$L = \sqrt{\frac{\alpha}{4D}} \frac{(x_0^2 + x^2) \cosh \sqrt{4\alpha D} \, (t - t_0) - 2x_0 x}{\sinh \sqrt{4\alpha D} \, (t - t_0)}$$

$$+ \int_{t_0}^{t} \left[ \frac{1}{4D} \left( \frac{dy}{d\tau} \right)^2 + \alpha y^2 \right] d\tau . \tag{6.7}$$

Consequently, the propagator (4.5) has the form

$$G_A(x, t \mid x_0, t_0) = f(t - t_0)$$

$$\times \exp \left\{ -\sqrt{\frac{\alpha}{4D}} \frac{(x_0^2 + x^2) \cosh \sqrt{4\alpha D} \, (t - t_0) - 2x_0 x}{\sinh \sqrt{4\alpha D} \, (t - t_0)} \right\} , \tag{6.8}$$

where

$$f(t - t_0) \equiv \int_{0,t_0}^{0,t} \exp\left\{ -\int_{t_0}^{t} \left[ \frac{1}{4D}\left(\frac{dy}{d\tau}\right)^2 + \alpha y^2 \right] \right\} d[y(\tau)] . \quad (6.9)$$

The remarkable fact is that the last path integral is over all paths $y(\tau)$ with $y(t_0) = y(t) = 0$ and is, therefore, independent of $x_0$ and $x$.

(c) The last step is the explicit calculation of the multiplicative factor $f(t - t_0)$. Again this can be done in many ways; a somewhat heuristic way is the following. As the paths $y(\tau)$ go from 0 at $\tau = t_0$ to 0 at $\tau = t$ they can be represented by a Fourier series with real expansion coefficients $a_n$

$$y(\tau) = \sum_{n=1}^{\infty} a_n \sin n\pi \frac{\tau - t_0}{t - t_0} . \quad (6.10)$$

Substitution gives

$$\int_{t_0}^{t} \left[ \frac{1}{4D}\left(\frac{dy}{d\tau}\right)^2 + \alpha y^2 \right] d\tau = \frac{t - t_0}{2} \sum_{n=1}^{\infty} \left[ \frac{1}{4D}\left(\frac{n\pi}{t - t_0}\right)^2 + \alpha \right] a_n^2 . \quad (6.11)$$

Also, the paths in the path integral (6.9) can be characterized by the expansion coefficients $a_n$ instead of the values of $y$ at the many intermediate times. Although the Jacobian $J$ of this transformation is not yet known to us we know that $J$ is independent of $\alpha$. Hence the function $f(t - t_0)$ must have the form

$$f(t - t_0) = (\text{factor}) \prod_{n=1}^{\infty} \int_{-\infty}^{+\infty} da_n \exp\left\{ -\frac{1}{2}(t - t_0) \right.$$

$$\left. \times \left[ \frac{1}{4D}\left(\frac{n\pi}{t - t_0}\right)^2 + \alpha \right] a_n^2 \right\}, \quad (6.12)$$

where the quantity denoted (factor) does not depend on $\alpha$. As the integral equals

$$\left(\frac{2\pi}{t - t_0}\right)^{1/2}\left\{ \frac{1}{4D}\left(\frac{n\pi}{t - t_0}\right)^2 + \alpha \right\}^{-1/2},$$

one can bring another $\alpha$-independent factor outside of the infinite product and finds

$$f(t - t_0) = F \prod_{n=1}^{\infty} \left\{ 1 + 4\alpha D \left(\frac{t - t_0}{n\pi}\right)^2 \right\}^{-1/2}, \quad (6.13)$$

where $F$ does not depend on $\alpha$. The infinite product has the value

$$\left\{ \frac{\sinh (t - t_0) \sqrt{4\alpha D}}{(t - t_0) \sqrt{4\alpha D}} \right\}^{-1/2},$$

according to Eq. (4.5.68) of [55], so we have

$$f(t - t_0) = F \left\{ \frac{\sinh \sqrt{4\alpha D} \ (t - t_0)}{\sqrt{4\alpha D} \ (t - t_0)} \right\}^{-1/2}. \tag{6.14}$$

The factor $F$, which is still unknown, now simply follows by taking the limit $\alpha \to 0$. In this limit the path integral (6.9) simplifies to (3.1) which we calculated in Section 1.3 and which equals $\{4\pi D(t-t_0)\}^{-1/2}$. As the bracketed expression tends to unity we find $F = \{4\pi D(t-t_0)\}^{-1/2}$. Collecting all results the propagator is found to equal

$$G_A(x, t \mid x_0, t_0) = \left\{ \pi \sqrt{\frac{4D}{\alpha}} \sinh \sqrt{4\alpha D} \ (t - t_0) \right\}^{-1/2}$$

$$\times \exp \left\{ -\sqrt{\frac{\alpha}{4D}} \ \frac{(x_0^2 + x^2) \cosh \sqrt{4\alpha D} \ (t - t_0) - 2x_0 x}{\sinh \sqrt{4\alpha D} \ (t - t_0)} \right\}. \tag{6.15}$$

The method of this section enables one to find an explicit expression for any Wiener integral with an exponential which is a quadratic form. A variety of physical applications is discussed in [34]. The reader should also consult the recent review by Khandekar and Lawande [56].

## References

[1]   V. Volterra, *Theory of Functionals and of Integral and Integrodifferential Equations* (McGraw-Hill, New York, 1965).
[2]   P.J. Daniell, *Ann. Math.* **19** (1918) 279.
[3]   P.J. Daniell, *Ann. Math.* **20** (1918) 1.
[4]   P.J. Daniell, *Ann. Math.* **20** (1919) 281.
[5]   P.J. Daniell, *Ann. Math.* **21** (1920) 203.
[6]   M. Kac, *Bull. Am. Math. Soc.* **72**, Part II (1966) 52.
[7]   N. Wiener, *Proc. Nat. Acad. Sci. USA* **7** (1921) 253.
[8]   N. Wiener, *Proc. Nat. Acad. Sci. USA* **7** (1921) 294.
[9]   N. Wiener, *J. Math. Phys.* MIT **2** (1923) 131.
[10]  N. Wiener, *Proc. Lond. Math. Soc.* **22** (1924) 454.
[11]  N. Wiener, *Acta Math.* **55** (1930) 117, Section 13; reprinted in Ref. 54.
[12]  G.J. Papadopoulos in *Path Integrals*, G.J. Papadopoulos and J.T. Devreese, eds. (Plenum, New York, 1978) p. 85.
[13]  G.J. Papadopoulos and J.T. Devreese, eds., *Path Integrals* (Plenum, New York, 1978).
[14]  J. Kirkwood, *Phys. Rev.* **44** (1933) 31.

[15] P.A.M. Dirac, *Phys. Zeits der Sowjetunion* **3** (1932) 64 (reprinted in Ref. 16, p. 312).

[16] J. Schwinger, ed., *Selected Papers on Quantum Electrodynamics* (Dover, New York, 1958).

[17] R.P. Feynman, "The Principle of Least Action in Quantum Mechanics," Ph.D. Thesis, Princeton University (1942), unpublished.

[18] R.P. Feynman, *Rev. Mod. Phys.* **20** (1948) 367 (reprinted in Ref. 16, p. 321).

[19] M. Kac, *Probability and Related Topics in Physical Sciences* (Interscience, New York, 1959) Chap. IV.

[20] R.P. Feynman, *Phys. Rev.* **90** (1953) 1116.

[21] R.P. Feynman, *Phys. Rev.* **91** (1953) 1291.

[22] R.P. Feynman, *Phys. Rev.* **91** (1953) 1301.

[23] R.P. Feynman, *Phys. Rev.* **94** (1954) 262.

[24] D. ter Haar, *Phys. Rev.* **95** (1954) 895.

[25] R. Kikuchi, *Phys. Rev.* **96** (1954) 563.

[26] R. Kikuchi, *Phys. Rev.* **99** (1955) 1684.

[27] R. Kikuchi, H.H. Denman and C.L. Schreiber, *Phys. Rev.* **119** (1960) 1823.

[28] C.E. Hecht, R. Kikuchi and P.R. Stein, *Phys. Rev.* **131** (1963) 907.

[29] I.M. Gelfand and A.M. Yaglom, *J. Math. Phys.* **1** (1960) 48.

[30] S.G. Brush, *Rev. Mod. Phys.* **33** (1961) 79.

[31] R.P. Feynman, *Phys. Rev.* **80** (1950) 440.

[32] S.F. Edwards and R.E. Peierls, *Proc. Roy. Soc.* **A22** (1954) 424.

[33] E. Hopf, *J. Mech. and Rat. Anal.* **1** (1952) 87.

[34] R.P. Feynman and A.R. Hibbs, *Quantum Mechanics and Path Integrals* (McGraw-Hill, New York, 1965).

[35] L.S. Schulman, *Techniques and Applications of Path Integration* (Wiley, New York, 1981).

[36] C. Garrod, *Rev. Mod. Phys.* **38** (1966) 483.

[37] F.W. Wiegel, *Phys. Reports* **16** (1975) 57.

[38] A. Neveu, *Rep. Prog. Phys.* **40** (1977) 599.

[39] R. Dashen, *J. Math. Phys.* **20** (1979) 894.

[40] C. DeWitt-Morette, A. Maheshwari and B. Nelson, *Phys. Reports* **50** (1979) 255.

[41] A.M. Arthurs, ed., *Functional Integration and its Applications* (Clarendon Press, Oxford, 1975).

[42] S. Albeverio, P. Combe and R. Hoegh-Krohn, eds., *Feynman Path Integrals*, Lecture Notes in Physics **106** (Springer, Heidelberg, 1979).

[43] A. Einstein, *Investigations on the Theory of the Brownian Movement* (Dover, New York, 1956).

[44] S. Chandrasekhar, *Rev. Mod. Phys.* **15** (1943) 1 (reprinted in Ref. 45).

[45] N. Wax, ed., *Selected Papers on Noise and Stochastic Processes* (Dover, New York, 1954).

[46] M.N. Barber and B.W. Ninham, *Random and Restricted Walks* (Gordon and Breach, New York, 1970).

[47] H.C. Berg and E.M. Purcell, *Biophys. J.* **20** (1977) 193.

[48] C. DeLisi and F.W. Wiegel, *Proc. Nat. Acad. Sci. USA* **78** (1981) 5569.

[49] A.S. Perelson, C. DeLisi and F.W. Wiegel, eds., *Cell Surface Dynamics: Concepts and Models* (Marcel Dekker, New York, 1984).

[50]  F.W. Wiegel, "Diffusion of Proteins in Membranes," in *Cell Surface Dynamics: Concepts and Models*, A.S. Perelson, C. DeLisi and F.W. Wiegel, eds. (Marcel Dekker, New York, 1984) p. 135.

[51]  F.W. Wiegel, *Phys. Reports* **95** (1983) 283.

[52]  B.B. Mandelbrot, *Fractals* (Freeman and Company, San Francisco, 1977).

[53]  B.B. Mandelbrot, *The Fractal Geometry of Nature* (Freeman and Company, San Francisco, 1982).

[54]  N. Wiener, *Generalized Harmonic Analysis and Tauberian Theorems* (MIT Press, Cambridge, Mass., 1964).

[55]  A. Abramowitz and I.A. Stegun, *Handbook of Mathematical Functions* (Dover, New York, 1970).

[56]  D.C. Khandekar and S.V. Lawande, *Phys. Reports* **137** (1986) 115.

# II. BROWNIAN MOTION IN A FIELD OF FORCE

## 2.1 Path-integral representation of the propagator

In the previous chapter the reader was introduced to several important concepts in path integration. In order to consolidate his or her knowledge this chapter is devoted to the application of these ideas to a simple, but physically important system: a particle which performs Brownian motion in a fluid, but which is at the same time subject to an external force. The aim of the chapter is to describe the system in terms of path integrals, and to extract some physically interesting insights from this description.

Let $\mathbf{r} = (x, y, z)$ denote the three Cartesian coordinates, let $\mathbf{F}(\mathbf{r})$ denote the external force, and let $c(\mathbf{r}, t)\, d\mathbf{r}$ denote the number of Brownian particles in a small volume element $d\mathbf{r} = dx\, dy\, dz$ around position $\mathbf{r}$. The first step in the development consists of deriving a partial differential equation for the concentration $c(\mathbf{r}, t)$. The equation has the form

$$\frac{\partial c}{\partial t} = - \operatorname{div} \mathbf{j} \, , \qquad (1.1)$$

where the particle current density $\mathbf{j}$ now consists of two terms

$$\mathbf{j} = -D\nabla c + \frac{c\mathbf{F}}{f} \ . \tag{1.2}$$

The first term which has components $(-D\partial c/\partial x, \ -D\partial c/\partial y, \ -D\partial c/\partial z)$, is due to the Brownian diffusion of the particles and is the generalization of (I.2.1). The second term is valid only if the friction coefficient $f$ of the Brownian particles is "large," in which case an external force $\mathbf{F}$ imparts a drift velocity

$$\mathbf{v} = \frac{\mathbf{F}}{f} \tag{1.3}$$

to the Brownian particles, which results in a drift current density $+c\mathbf{F}/f$. Combination of these equations gives

$$\frac{\partial c}{\partial t} = D\triangle c - f^{-1} \operatorname{div} (c\mathbf{F}) \ , \tag{1.4}$$

where $\triangle = \partial^2/\partial x^2 + \partial^2/\partial y^2 + \partial^2/\partial z^2$ is the Laplacian operator.

The second step consists in finding a path-integral representation for the propagator $G$ of this equation, which is the solution of (1.4) that vanishes at infinity and approaches a delta function $\delta(\mathbf{r} - \mathbf{r}_0)$ for $t \downarrow t_0$. The one-dimensional case of this problem has given rise to an extensive literature; for the record only we list most of the relevant references [1–23]. The most straightforward way to derive a path integral is to write (denoting Boltzmann's constant by $k_B$ and the absolute temperature by $T$)

$$c(\mathbf{r}, \ t) = p(\mathbf{r}, \ t) \exp\left( + \frac{1}{2k_BT} \int_{\mathbf{r}_0}^{\mathbf{r}} \mathbf{F} \cdot d\mathbf{r} \right) \ . \tag{1.5}$$

In this formula the line integral follows any continuous contour which starts at $\mathbf{r}_0$ and ends at $\mathbf{r}$. The value of the line integral will be independent of the form of this contour provided the external force field $\mathbf{F}$ is conservative, in which case $\mathbf{F}$ can be written as the gradient of a scalar function

$$\mathbf{F} = -\nabla\phi \ . \tag{1.6}$$

For the time being we shall assume that this is the case. When (1.5) is substituted into (1.4) one finds that $p$ must be the solution of

$$\frac{\partial p}{\partial t} = D\triangle p - V(\mathbf{r})p \ , \tag{1.7a}$$

where

$$V \equiv \frac{\mathbf{F}^2}{4k_{\mathrm{B}}Tf} + \frac{\mathrm{div}\mathbf{F}}{2f} , \tag{1.7b}$$

and where we used the Einstein relation which connects the diffusion coefficient, the friction coefficient $f$ and Boltzmann's constant

$$Df = k_{\mathrm{B}}T . \tag{1.8}$$

Moreover, in the limit $t \downarrow t_0$ the function $p$ has to approach $\delta(\mathbf{r} - \mathbf{r}_0)$. Hence $p$ is given by the three-dimensional generalization of (I.4.5) and one finds, for $t > t_0$, the propagator

$$G(\mathbf{r}, t \mid \mathbf{r}_0, t_0) = \exp\left(\frac{1}{2k_{\mathrm{B}}T}\int_{\mathbf{r}_0}^{\mathbf{r}} \mathbf{F} \cdot d\mathbf{r}\right)$$

$$\times \int_{\mathbf{r}_0, t_0}^{\mathbf{r}, t} \exp\left\{-\frac{1}{4D}\int_{t_0}^{t}\left(\frac{d\mathbf{r}}{d\tau}\right)^2 d\tau - \int_{t_0}^{t} V\, d\tau\right\} d[\mathbf{r}(\tau)] . \tag{1.9}$$

Before we consider the physical interpretation of this result it should be pointed out that Hunt and Ross [24] have shown that this form of the path integral representation is also valid in the case in which the external force is non-conservative (also cf. Ref. 32). In this chapter we shall, however, restrict ourselves to the case of a conservative external force.

## 2.2.  Physical interpretation: the limit $D \to 0$

The propagator $G(\mathbf{r}, t \mid \mathbf{r}_0, t_0)$ is the probability density to find the Brownian particle in $\mathbf{r}$ at time $t$ if it was released in $\mathbf{r}_0$ at time $t_0$. Hence the quantity

$$W[\mathbf{r}(\tau)] \equiv \exp\left\{\frac{1}{2k_{\mathrm{B}}T}\int_{\mathbf{r}_0}^{\mathbf{r}} \mathbf{F} \cdot d\mathbf{r} - \frac{1}{4D}\int_{t_0}^{t}\left(\frac{d\mathbf{r}}{d\tau}\right)^2 d\tau - \int_{t_0}^{t} V\, d\tau\right\} \tag{2.1}$$

equals the probability-density functional for the particle to follow a specific trajectory $\mathbf{r}(\tau)$. Writing

$$\int_{\mathbf{r}_0}^{\mathbf{r}} \mathbf{F} \cdot d\mathbf{r} = \int_{t_0}^{t} \mathbf{F} \cdot \frac{d\mathbf{r}}{d\tau} d\tau$$

and using the explicit form (1.7b) one finds the alternative and very suggestive expression

$$W[\mathbf{r}(\tau)] = \exp\left\{ -\frac{1}{4D} \int_{t_0}^{t} L[\mathbf{r}(\tau)]\, d\tau \right\} \tag{2.2a}$$

$$L[\mathbf{r}(\tau)] = \left[ \frac{d\mathbf{r}}{d\tau} - \frac{\mathbf{F}}{f} \right]^2 + 2\frac{D}{f}\, \text{div}\, \mathbf{F} . \tag{2.2b}$$

In order to extract some physical insights from the theory it is natural first to consider the limit $D \to 0$, in which the fluctuations due to Brownian motion are switched off. In this limit the second term on the right-hand side of (2.2b) vanishes and, as the coefficient $(4D)^{-1}$ is infinite, (2.2a) shows that the only contributing trajectory is the solution of

$$\int_{t_0}^{t} \left[ \frac{d\mathbf{r}}{d\tau} - \frac{\mathbf{F}}{f} \right]^2 d\tau = 0 , \tag{2.3}$$

that is, the solution of

$$\frac{d\mathbf{r}}{d\tau} = \frac{\mathbf{F}}{f} . \tag{2.4}$$

But this is exactly the equation of motion of a particle in an external field of force $\mathbf{F}$ in the presence of very large friction with the medium. Hence, in the limit $D \to 0$ the path integral leads us back to the dynamical equation of "ordinary" classical mechanics (this was noticed for the first time by the author, cf. Refs. 1, 2).

For a finite value of the diffusion coefficient $D$ all trajectories $\mathbf{r}(\tau)$ will contribute to the path integral (1.9). The largest contribution will come from the trajectory for which the integral in the exponential in (2.2a) is as small as possible. According to the calculus of variations [25] the stationarity of the integral

$$\delta \int_{t_0}^{t} L\, d\tau = 0 , \qquad (\mathbf{r}(t_0) = \mathbf{r}_0,\ \mathbf{r}(t) = \mathbf{r}) , \tag{2.5}$$

implies the Euler-Lagrange equations

$$\frac{\partial L}{\partial x} - \frac{d}{d\tau}\frac{\partial L}{\partial \dot{x}} = 0 , \tag{2.6a}$$

$$\frac{\partial L}{\partial y} - \frac{d}{d\tau}\frac{\partial L}{\partial \dot{y}} = 0 , \tag{2.6b}$$

$$\frac{\partial L}{\partial z} - \frac{d}{d\tau}\frac{\partial L}{\partial \dot{z}} = 0 , \tag{2.6c}$$

where the dot denotes the time derivative. Substituting the explicit form (2.2b) one finds that the most probable trajectory $(x_c(\tau), y_c(\tau), z_c(\tau))$ is the solution of a set of three coupled second order ordinary differential equations

$$\frac{d^2x_c}{d\tau^2} = 2D\frac{\partial V}{\partial x_c} + \frac{\dot{y}_c}{f}\left(\frac{\partial F_1}{\partial y_c} - \frac{\partial F_2}{\partial x_c}\right) + \frac{\dot{z}_c}{f}\left(\frac{\partial F_1}{\partial z_c} - \frac{\partial F_3}{\partial x_c}\right), \quad (2.7a)$$

$$\frac{d^2y_c}{d\tau^2} = 2D\frac{\partial V}{\partial y_c} + \frac{\dot{z}_c}{f}\left(\frac{\partial F_2}{\partial z_c} - \frac{\partial F_3}{\partial y_c}\right) + \frac{\dot{x}_c}{f}\left(\frac{\partial F_2}{\partial x_c} - \frac{\partial F_1}{\partial y_c}\right), \quad (2.7b)$$

$$\frac{d^2z_c}{d\tau^2} = 2D\frac{\partial V}{\partial z_c} + \frac{\dot{x}_c}{f}\left(\frac{\partial F_3}{\partial x_c} - \frac{\partial F_1}{\partial z_c}\right) + \frac{\dot{y}_c}{f}\left(\frac{\partial F_3}{\partial y_c} - \frac{\partial F_2}{\partial z_c}\right), \quad (2.7c)$$

Now these equations are *not* identical to the classical equations of motion of a fictitious particle of unit mass, moving in an external field of force with scalar potential

$$\phi = -2DV = -\frac{\mathbf{F}^2}{2f^2} - \frac{D}{f}\,\text{div}\,\mathbf{F} . \quad (2.8)$$

The trajectory of the fictitious particle must be solved from (2.7) under the boundary conditions

$$\mathbf{r}_c(t_0) = \mathbf{r}_0 , \qquad \mathbf{r}_c(t) = \mathbf{r} . \quad (2.9)$$

Hence, for small values of the diffusion coefficient, an asymptotic expression for the propagator is simply given by

$$G(\mathbf{r}, t \mid \mathbf{r}_0, t_0) \cong K(t)\exp\left\{-\frac{1}{4D}\int_{t_0}^{t} L\,[\mathbf{r}_c(\tau)]\,d\tau\right\} \quad (D \downarrow 0) , \quad (2.10)$$

where the value of $K$ can be determined from the requirement that the integral of $G$ over all of $\mathbf{r}$ space should be equal to unity for all times $t$.

A simple example to illustrate the formalism has been discussed by Wiegel and Ross [32]. It consists of Brownian motion in a plane, in the special case in which (2.4) reads

$$\dot{x} = -\omega y , \qquad \dot{y} = +\omega x . \quad (2.11)$$

Hence the external field has components $F_1 = -\omega f y$, $F_2 = +\omega f x$ and as $\partial F_1/\partial y \neq \partial F_2/\partial x$ this is a non-conservative force field. In the absence of fluctuations ($D = 0$) the trajectory of the system is the circle $x^2 + y^2 = x^2(0) + y^2(0)$ through the initial position $x(0)$, $y(0)$. This trajectory is traversed with constant angular velocity $\omega$. If $x(t)$ and $y(t)$ are interpreted as the concentrations of two substances (for example two chemical reagents, two ecological species, or an antigen-antibody pair in the immune system [33]) this circular motion is the simplest example of the limit cycles which often occur in such systems. In the presence of Brownian motion the potential (1.7b) has the form

$$V = \frac{\omega^2}{4D} (x^2 + y^2) , \qquad (2.12)$$

and the equations of motion (2.7a, b) read

$$\frac{d^2x_c}{d\tau^2} = \omega^2 x_c - 2\omega \frac{dy_c}{d\tau} , \qquad (2.13a)$$

$$\frac{d^2y_c}{d\tau^2} = \omega^2 y_c + 2\omega \frac{dx_c}{d\tau} . \qquad (2.13b)$$

The general solution of these equations is a spiral

$$x_c(\tau) = r_0 \cos (\omega\tau + \phi_0) + s_0\tau \cos (\omega\tau + \psi_0) , \qquad (2.14a)$$

$$y_c(\tau) = r_0 \sin (\omega\tau + \phi_0) + s_0\tau \sin (\omega\tau + \psi_0) , \qquad (2.14b)$$

in which the constants $r_0, s_0, \phi_0, \psi_0$ are determined by the initial position and velocity at $t_0 = 0$, in the following way. For $\tau = 0$, one has

$$x_c(0) = r_0 \cos \phi_0 , \qquad y_c(0) = r_0 \sin \phi_0 , \qquad (2.15)$$

so $r_0$ and $\phi_0$ are the polar coordinates of the initial position in the $x$, $y$ plane. Also, for the velocity one finds

$$\dot{x}_c(0) = -\omega y_c(0) + s_0 \cos \psi_0 , \qquad \dot{y}_c(0) = \omega x_c(0) + s_0 \sin \psi_0 , \qquad (2.16)$$

and, hence, $s_0$ and $\psi_0$ are the polar coordinates (in velocity space) of the difference between the initial velocity and that initial velocity which would lead to circular motion. The extremal value of the Lagrangian is found upon substitution of (2.14a, b) into (2.2b) and integration over $\tau$; this gives

$$\int_0^t L[\mathbf{r}_c(\tau)] \, d\tau = \frac{s_0^2 t}{4D} . \qquad (2.17)$$

Of course, $s_0$ is determined by the requirement that the trajectory passes through $\mathbf{r}$ at final time $t$. This gives

$$x = r_0 \cos (\omega t + \phi_0) + s_0 t \cos (\omega t + \psi_0) \, , \tag{2.18a}$$

$$y = r_0 \sin (\omega t + \phi_0) + s_0 t \sin (\omega t + \psi_0) \, , \tag{2.18b}$$

from which one finds

$$s_0^2 t^2 = \{ x - r_0 \cos (\omega t + \phi_0)\}^2 + \{ y - r_0 \sin (\omega t + \phi_0)\}^2 \, . \tag{2.19}$$

Substitution of (2.17) and the last formula into (2.10) gives

$$G(x, y, t \mid x_0, y_0) \cong K(t) \exp \left[ - \frac{1}{4Dt} \{x - r_0 \cos (\omega t + \phi_0)\}^2 \right.$$

$$\left. - \frac{1}{4Dt} \{y - r_0 \sin (\omega t + \phi_0)\}^2 \right] \, . \tag{2.20}$$

If the function $K(t)$ is determined by the requirement that at time $t$ the integral of $G$ over all $x$ and $y$—which is the total probability to find the system somewhere—should equal unity one finds

$$K(t) = (4\pi Dt)^{-1} \, . \tag{2.21}$$

It was noticed in Ref. 32 that the last two equations actually give the exact result for the propagator. This is the case because for the choice (2.11) the Lagrangian is quadratic in the variables $x$, $y$, $\dot{x}$ and $\dot{y}$. We shall study the quadratic approximation of an arbitrary Lagrangian in the next section.

## 2.3. The quadratic approximation

The next approximation of the propagator which is suggested by the path-integral representation (1.9) is generally known as the quadratic approximation. The approximation was worked out for Brownian motion in a field of force by the author in the late sixties [2] but is being rediscovered independently by other authors every couple of years [26 – 28]. The results have been applied by Freed [29] to polymer-chain statistics, by Laing and Freed [30] to the semiclassical theory of scattering, and by Kitahara, Metiu and Ross [31] to the theory of nucleation. Following Ref. 2 we shall demonstrate the method for the one-dimensional version of (1.9) which, apart from the first factor, is identical to (I.4.5):

$$G_V(x, t) \equiv \int_{x_0, t_0}^{x, t} \exp \left\{ - \frac{1}{4D} \int_{t_0}^{t} \left( \frac{dx}{d\tau} \right)^2 d\tau - \int_{t_0}^{t} V(x) \, d\tau \right\} d[x(\tau)] \, . \tag{3.1}$$

The method discussed here forms an extension of the method of Section 1.6.

Write

$$x(\tau) = x_c(\tau) + \delta(\tau) \ , \tag{3.2}$$

where $x_c(\tau)$ stands for the most probable trajectory solved from (2.7), and where $\delta(\tau)$ denotes the deviation of a path from the most probable path. Hence one has $\delta(t_0) = 0$, $\delta(t) = 0$. Expand the exponential in (3.1) in powers of $\delta$. The term of order 0 gives rise to a factor

$$\exp\left\{ -\frac{1}{4D}\int_{t_0}^{t}\left(\frac{dx_c}{d\tau}\right)^2 d\tau - \int_{t_0}^{t} V(x_c)\, d\tau \right\} \tag{3.3}$$

in front of the path integral. The linear term gives no contribution because of the definition of $x_c(\tau)$. The second order term is of the form

$$-\frac{1}{4D}\int_{t_0}^{t}\left(\frac{d\delta}{d\tau}\right)^2 d\tau - \frac{1}{2}\int_{t_0}^{t} V_c''(\tau)\, \delta^2(\tau)\, d\tau \ , \tag{3.4}$$

where $V_c''(\tau)$ denotes the function $V''(x_c(\tau))$. The quadratic approximation consists of neglecting all terms of higher order than the second. It will be a good approximation if the path probability decreases rapidly with increasing variation of the path form the optimal one, i.e., if $D$ is small.

Collecting the results found till now, and using the definitions (I.2.11) and (I.4.6), we find the expression

$$G_V(x, t) \cong F(x, t) \exp\left\{ -\frac{1}{4D}\int_{t_0}^{t}\left(\frac{dx_c}{d\tau}\right)^2 d\tau - \int_{t_0}^{t} V(x_c)\, d\tau \right\} \ , \tag{3.5}$$

$$F(x, t) = \lim (4\pi\varepsilon D)^{-(N+1)/2}\int_{-\infty}^{\infty} d\delta_1 \int_{-\infty}^{\infty} d\delta_2 \dots \int_{-\infty}^{\infty} d\delta_N$$

$$\times \exp\left\{ -\frac{1}{4D\varepsilon}\sum_{j=0}^{N}(\delta_{j+1} - \delta_j)^2 - \frac{1}{2}\varepsilon\sum_{j=1}^{N} V_j'' \delta_j^2 \right\} \ , \tag{3.6}$$

where $V_j'' = V''(x_c(\tau_j))$, $\delta_0 = \delta_{N+1} = 0$. The multiple integral can be evaluated with the method of Section 1.3. Write the exponential in (3.6) in the form $-(4D\varepsilon)^{-1}\sum_{k,l=1}^{N} \delta_k B_{k,l} \delta_l$. The $N \times N$ matrix $B$ has zero matrix elements apart from those in the main diagonal and the two neighboring diagonals

$$B_{k,k} = 2 + 2D\varepsilon^2 V_k'' \ , \tag{3.7a}$$

$$B_{k,k+1} = B_{k,k-1} = -1 \ . \tag{3.7b}$$

Calling its eigenvalues $\lambda_j$ one finds

$$F(x, t) = \lim (4\pi\varepsilon D)^{-(N+1)/2} \prod_{j=1}^{N} (4\pi\varepsilon D/\lambda_j)^{1/2}$$

$$= \lim (4\pi\varepsilon D \det B_N)^{-1/2} . \tag{3.8}$$

If the determinant $\det B_N$ of the matrix $B$ is expanded in the elements of its last column the following relation is obtained

$$\det B_N = (2 + 2\varepsilon^2 DV''_N) \det B_{N-1} - \det B_{N-2} . \tag{3.9}$$

In the absence of an external field we found in Section 1.3 that the determinant equals $N + 1$. One therefore introduces the quantity

$$C_N = \frac{\det B_N}{N + 1} , \tag{3.10}$$

which obeys the recurrent relation

$$C_N - 2C_{N-1} + C_{N-2} = \frac{-2}{N+1}(C_{N-1} - C_{N-2}) + 2\varepsilon^2 D\frac{NV''_N}{N+1}C_{N-1} . \tag{3.11}$$

In the limit $N \to \infty$, $\varepsilon \to 0$, $(N + 1)\varepsilon = t - t_0$ the discrete variable $C_N$ tends to a continuous function $C(x, t)$ which obeys the ordinary differential equation

$$\frac{d^2C}{dt^2} + \frac{2}{(t - t_0)} \frac{dC}{dt} = 2DV'' (x_c (t))C . \tag{3.12}$$

The initial conditions are

$$C = 1, \quad \frac{dC}{dt} = 0 , \quad \text{at } t = t_0 . \tag{3.13}$$

Substitution of these results into (3.8) gives

$$F(x, t) = \{4\pi D(t - t_0) C(x, t)\}^{-1/2} . \tag{3.14}$$

The last step in our evaluation of the quadratic approximation of the path integral consists of the explicit calculation of the function $C(x, t)$. For a given, fixed value of $x$ the last formula suggests that one introduce

$$H(t) \equiv (t - t_0) C(x, t) \tag{3.15}$$

as the unknown function, which is found to obey

$$\frac{d^2H}{dt^2} = 2DV'' (x_c(t))H \tag{3.16}$$

with boundary conditions

$$H = 0, \quad \frac{dH}{dt} = 1 \quad \text{at } t = t_0 . \tag{3.17}$$

This equation is most easily solved if one chooses $x_c$ as the independent variable, instead of $t$. Indicating differentiation with respect to the new independent variable by a prime, and using (2.7) to find

$$\frac{dx_c}{dt} = (2E + 4DV)^{1/2} , \tag{3.18}$$

where $E$ is the "energy" of the fictitious particle, one has the equation

$$(2E + 4DV)H'' + 2DV'H' = 2DV''H , \tag{3.19}$$

with the initial conditions

$$H = 0, \quad H' = (2E + 4DV)^{-1/2} \quad \text{for} \quad x_c = x_0 . \tag{3.20}$$

As (3.19) can actually be written as

$$\{(2E + 4DV)H\}'' = \{6DV'H\}' , \tag{3.21}$$

the solution is

$$H(x_c) = (2E + 4DV(x_0))^{1/2} (2E + 4DV(x_c))^{1/2}$$

$$\times \int_{x_0}^{x_c} (2E + 4DV(y))^{-3/2} \, dy , \tag{3.22}$$

in the case $x_c > x_0$. Substitution of this formula and (3.14, 15) into (3.5) gives, for the quadratic approximation of the path integral (3.1),

$$G_V(x, t) \cong \left\{ 4\pi D(2E + 4DV(x_0))^{1/2} (2E + 4DV(x))^{1/2} \right.$$

$$\left. \times \int_{x_0}^{x} (2E + 4DV(y))^{-3/2} \, dy \right\}^{-1/2}$$

$$\times \exp \left\{ -\frac{1}{4D} \int_{t_0}^{t} \left( \frac{dx_c}{d\tau} \right)^2 d\tau - \int_{t_0}^{t} V(x_c) \, d\tau \right\} . \tag{3.23}$$

This formula should of course be exact when $V(x)$ is at most of second order in $x$, since then no terms occur of higher order than second.

An important application of these results is to the Ornstein-Uhlenbeck process (cf. Ref. I-44, I-45). This is the one-dimensional Brownian motion of a particle subject to an harmonic force

$$F(x) = -k_0 x . \tag{3.24}$$

In this case the function $V(x)$ of Eq. (1.7b) is given by

$$V(x) = \frac{\gamma^2 x^2}{4D} - \frac{\gamma}{2} , \tag{3.25}$$

where

$$\gamma \equiv \frac{k_0}{f} . \tag{3.26}$$

In this simple case the optimal trajectory $x_c(\tau)$ can be solved from (2.7) in a straightforward way and has the form

$$x_c(\tau) = x_0 e^{-\gamma\tau} + A(e^{\gamma\tau} - e^{-\gamma\tau}) , \tag{3.27a}$$

$$A = (e^{\gamma t} - e^{-\gamma t})^{-1} (x - x_0 e^{-\gamma t}) . \tag{3.27b}$$

Substitution gives, for the two main exponentials in (1.9) and in (3.23),

$$\frac{1}{2k_B T} \int_{x_0}^{x} F(x') \, dx' = \frac{\gamma}{4D} (x_0^2 - x^2) , \tag{3.28}$$

$$\int_0^t \left\{ \left(\frac{dx_c}{d\tau}\right)^2 + 4DV(x_c) \right\} d\tau = -2\gamma Dt + \gamma(1 - e^{-2\gamma t})^{-1} (x - x_0 e^{-\gamma t})^2$$
$$- \gamma(1 - e^{+2\gamma t})^{-1} (x - x_0 e^{+\gamma t})^2 . \tag{3.29}$$

The forefactor $F(x, t) = \{4\pi DH(x, t)\}^{-1/2}$ can most easily be calculated from (3.16, 17), which now reads

$$\frac{d^2 H}{dt^2} = \gamma^2 H , \tag{3.30}$$

with the solution

$$H(t) = \frac{1}{2\gamma} (e^{\gamma t} - e^{-\gamma t}) . \tag{3.31}$$

Combination of (3.28, 29, 31) and (3.23) gives, for the full propagator of the Ornstein-Uhlenbeck process,

$$G(x, t \,|\, x_0, 0) = \left\{ 2\pi \frac{D}{\gamma} (1 - e^{-2\gamma t}) \right\}^{-1/2} \exp \left\{ - \frac{\gamma}{2D} \frac{(x - x_0 e^{-\gamma t})^2}{1 - e^{-2\gamma t}} \right\} . \tag{3.32}$$

As $V(x)$ is quadratic this result of the quadratic approximation is rigorous in the present case. This fact essentially represents the principle of Onsager and Machlup from linear irreversible thermodynamics [34 – 36].

## 2.4.  Large versus small fluctuations: the hopping-paths approximation

The formula (3.23) which results from a straightforward application of the quadratic approximation (which itself is just a special case of the saddle-point method [37]) by no means exhausts the usefulness of this method. The quadratic approximation which leads to this formula, takes care of small fluctuations of the path of the particle around the most probable path. However, it would be erroneous to assume that (3.23) is a good approximation for all times $t - t_0$ provided $D$ is small. In fact, even if the diffusion coefficient is small the quadratic approximation will break down in the limit $t - t_0 \to \infty$ (unless $V$ is at most quadratic). To the author's knowledge this was pointed out for the first time by Jalickee, Wiegel and Vezzetti [38] and in some later publications by Wiegel [1−37, 39, 40] and by Moreau [26]. For large times one has to use an extension of the quadratic approximation which will be called the method of hopping paths.

Essentially, the method of hopping paths takes large fluctuations $\delta(\tau)$ into account by writing the propagator in the form

$$G_V(x, t) \cong C \sum_s \exp \left\{ - \frac{1}{4D} \int_{t_0}^{t} L[x_s(\tau)] \, d\tau \right\} , \qquad (4.1)$$

instead of the expression (3.5, 6). In this formula $C$ is a constant which usually will not be evaluated explicitly, and the summation extends over all those functions $x_s(\tau)$ which are approximate(but not exact!)solutions of the optimal path condition

$$\frac{\partial L}{\partial x} - \frac{d}{dt} \frac{\partial L}{\partial \dot{x}} = 0 . \qquad (4.2)$$

As for the present case $L = \dot{x}^2 + 4DV$ this is the differential equation

$$\ddot{x} = 2 DV' . \qquad (4.3)$$

An important case is a bistable potential for which the external force on the Brownian particle is given by

$$F(x) = - \alpha^2(x - a)(x - b)(x - c) , \qquad (4.4)$$

with $a < b < c$; see Fig. 2.1. Diffusion in this bistable potential has been studied recently with different methods by van Kampen [41], Dekker and van Kampen [42] and by Caroli, Caroli and Roulet [43]. In these papers van Kampen derives analytical results for an exactly solvable case, Dekker and van Kampen give a numerical solution and Caroli, Caroli and Roulet use the WKB method. In this section we shall outline the main steps in the hopping-paths approximation; for full details the reader is referred to Refs. 38 – 40 and I – 37. We shall use a potential $V(x)$ which corresponds

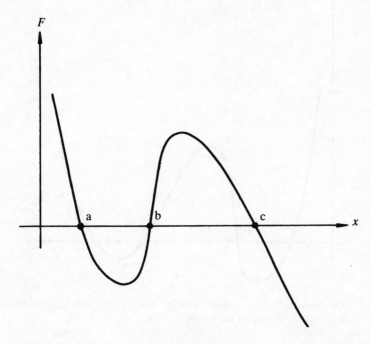

Fig. 2.1. A force field $F(x)$ in which the particle has two stable equilibrium points $x = a$ and $x = c$ and one unstable equilibrium point $x = b$.

to (4.4) and which will have two minima of slightly different depths, as indicated in Fig. 2.2. The essential steps are:

(a) The constant function $x_s(\tau) = x_a$ is a solution of (4.3). For such a solution the time integral of the "Lagrangian" which appears in the exponential of (4.1) equals

$$\int_{t_0}^{t} L[x_a] \, d\tau = 4DV(x_a) \, (t - t_0) \ . \qquad (4.5a)$$

(b) In the same way $x_s(\tau) = x_c$ is a solution of (4.3) which leads to an exponential

$$\int_{t_0}^{t} L[x_c] \, d\tau = 4DV(x_c) \, (t - t_0) \ . \qquad (4.5b)$$

(c) A typical "hopping-path" solution $x_s(\tau)$ has the form shown in Fig. 2.3. The value of $x_s$ hops from $x_a$ to $x_c$ and back at times $t_1, t_2, t_3, \ldots$

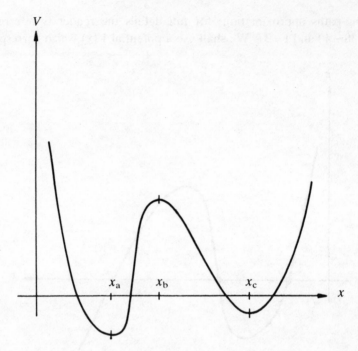

Fig. 2.2.   The "potential" $V(x)$, defined by (1.7b), where $F(x)$ is as in Fig. 2.1. Minima for $x = x_a$ and $x = x_c$, top of the barrier at $x = x_b$.

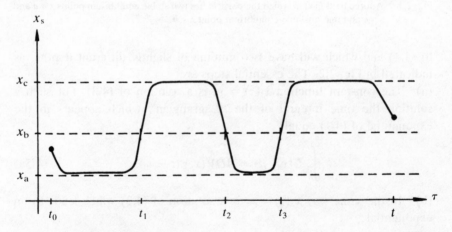

Fig. 2.3.   A typical "hopping-path fluctuation" $x_s(\tau)$, which dominates the path integral. The contributions of all these fluctuating paths are summed in Eqs. (4.1) and (4.8).

The contribution of a hopping path is determined by the last two expressions. However, for every transition between the two wells in the potential $V$ there is a correction factor $W$ due to the fact that $x_s$ is not constant during the transitions. To a good approximation one finds

$$W \cong K \exp\left(-\frac{\gamma}{D}\right) . \tag{4.6}$$

where $K$ is a constant and where, if $t_i$ is the time at which $x_s = b$,

$$\gamma \equiv \frac{1}{2} \int_{t_i}^{\infty} \{L[x_s(\tau)] - L[x_a]\} \, d\tau . \tag{4.7}$$

Now, upon substitution of the explicit form for $L$, and using for $x_s(\tau)$ the solution of (4.3) with the boundary conditions $x_s(t_i) = x_b$, $x_s(\infty) = x_a$, one finds that $\gamma$ is independent of $D$ for small values of $D$.

(d)   Hence, for small values of the diffusion coefficient the hopping-paths approximation (4.1) to the propagator is given by

$$G_V(x, t) \cong C \sum_{n=0}^{\infty} W^{2n} \int dt_1 \int dt_2 \ldots \int dt_{2n}$$

$$\times \exp\left[-V(x_a)\{(t_1 - t_0) + (t_3 - t_2) + \ldots + (t - t_{2n})\}\right.$$

$$\left. - V(x_c)\{(t_2 - t_1) + (t_4 - t_3) + \ldots + (t_{2n} - t_{2n-1})\}\right] . \tag{4.8}$$

In this formula we have assumed that $x_0$ and $x$ are both very close to $x_a$. The integrations are over the times at which the particle passes through $x_b$ and are subject to the constraint

$$t_0 < t_1 < t_2 < \ldots < t_{2n} < t . \tag{4.9}$$

(e)   The evaluation of the multiple integral in the summation in (4.8), which has the form of a repeated convolution product, proceeds by taking the Laplace transform of all functions with respect to the time differences in their arguments

$$\tilde{G}(s) \equiv \int_{t_0}^{\infty} G_V(x, t) \exp\{-s(t - t_0)\} \, dt , \tag{4.10}$$

$$\int_{t_0}^{\infty} \exp\{-V(x_a)(t_1 - t_0) - s(t_1 - t_0)\} dt_1 = \{V(x_a) + s\}^{-1} , \tag{4.11}$$

$$\int_{t_1}^{\infty} \exp\{-V(x_c)(t_2 - t_1) - s(t_2 - t_1)\} dt_2 = \{V(x_c) + s\}^{-1} . \tag{4.12}$$

Using the theorem that the Laplace transform of a convolution product of two functions equals the product of the Laplace transforms of the terms, one finds

$$\tilde{G}(s) \cong C \sum_{n=0}^{\infty} W^{2n} \{V(x_c) + s\}^{-n} \{V(x_a) + s\}^{-n-1}$$

$$= C \{V(x_c) + s\} [\{V(x_a) + s\} \{V(x_c) + s\} - W^2]^{-1} . \quad (4.13)$$

The propagator itself follows from the inverse Laplace transform

$$G_V(x, t) \cong (2\pi i)^{-1} \int_{s-i\infty}^{s+i\infty} \frac{C\{V(x_c) + s\}}{\{V(x_a) + s\} \{V(x_c) + s\} - W^2}$$

$$\times \exp \{+ s (t - t_0)\} \, ds . \quad (4.14)$$

The integrand has two poles at

$$s_{\pm} = -\frac{1}{2} \{V(x_a) + V(x_c)\} \pm \frac{1}{2} \sqrt{\{V(x_a) - V(x_c)\}^2 + 4W^2} . \quad (4.15)$$

When the contour of integration is deformed into two small circles around these poles, one finds

$$G_V(x, t) \cong C_+ \exp \{s_+(t - t_0)\} + C_- \exp \{s_-(t - t_0)\} , \quad (4.16)$$

where $C_+$ and $C_-$ are two constants whose values can be calculated with the aid of Cauchy's theorem. This essentially completes the calculation of the propagator in the hopping-paths approximation.

It should be noticed here that the long-time behavior of the propagator is determined by the larger of the two frequencies $s_+$ and $s_-$. This is $s_+$, which, for the special case in which the two minima in the potential have equal depths $(V(x_a) = V(x_c) \equiv V_0)$, gives the asymptotic form

$$G_V(x, t) \cong C_+ \exp \{(- V_0 + W) (t - t_0)\} , \qquad (t \to \infty) . \quad (4.17)$$

This can be interpreted as the product of a factor $\exp \{- V_0(t - t_0)\}$ which would arise if the Brownian particle simply stayed in the vicinity of $x_a$, times a factor $\exp \{+ W(t - t_0)\}$ due to penetration of the barrier around $x_b$. As $W$ has the non-analytic (in $D$) form (4.6) this is our first example of an approximate evaluation of a path integral which leads to a non-analytic result.

### References

[1]  F.W. Wiegel, *Physica* **33** (1967) 734.
[2]  F.W. Wiegel, *Physica* **37** (1967) 105.
[3]  R. Graham, *Springer Tracts Mod. Phys.* **66** (1973) 1.

[4]   R. Kubo, M. Matsuo and K. Kitahara, *J. Stat. Phys.* **9** (1973) 51.
[5]   W. Horsthemke and A. Bach, *Z. Phys.* **B22** (1975) 189.
[6]   H. Yahata, *Prog. Theor. Phys.* **52** (1974) 871.
[7]   K. Kitahara and H. Metiu, *J. Stat. Phys.* **15** (1976) 141.
[8]   H. Ueyama, *Physica* **84A** (1976) 392.
[9]   H. Ueyama, *Physica* **84A** (1976) 402.
[10]  H. Hasegawa, *Prog. Theor. Phys.* **55** (1976) 90.
[11]  H. Hasegawa, *Prog. Theor. Phys.* **56** (1976) 44.
[12]  B. Mühlschleger, Ref. I – 13, p. 39.
[13]  H. Haken, *Phys. Lett.* **55A** (1976) 323.
[14]  H. Haken, *Zeit. Phys.* **B24** (1976) 321.
[15]  H. Haken, *Synergetics*, 2nd ed. (Springer, Heidelberg, 1978) Section 6.6.
[16]  H. Dekker, *Physica* **84A** (1976) 205.
[17]  H. Dekker, *Physica* **85A** (1976) 363.
[18]  H. Dekker, *Physica* **92A** (1978) 438.
[19]  H. Dekker, *Phys. Lett.* **65A** (1978) 388.
[20]  H. Dekker, *Phys. Lett.* **67A** (1978) 90.
[21]  H. Dekker, *Phys. Rev.* **A19** (1979) 2102.
[22]  I.F. Bakhareva and A.A. Biryukov, *Zh. Fiz. Kim.* **48** (1974) 1985; *Russ. J. Phys. Chem.* **48** (1974) 1159.
[23]  B.C. Eu, *Physica* **90A** (1978) 288.
[24]  K.L.C. Hunt and J. Ross, *J. Chem. Phys.* **75** (1981) 976.
[25]  R. Courant and D. Hilbert, *Methods of Mathematical Physics,* Vol. 1 (Interscience, New York, 1953).
[26]  M. Moreau, *Physica* **90A** (1978) 410.
[27]  B.H. Lavenda, *Phys. Lett.* **71A** (1979) 304.
[28]  F. Langouche, D. Roekaerts and E. Tirapegui, *Physica* **97A** (1979) 195.
[29]  K.F. Freed, *J. Chem. Phys.* **54** (1971) 1453.
[30]  J.R. Laing and K.F. Freed, *Chem. Phys.* **19** (1977) 91.
[31]  K. Kitahara, H. Metiu and J. Ross, *J. Chem. Phys.* **63** (1975) 3156.
[32]  F. W. Wiegel and J. Ross, *Phys. Lett.* **84A** (1981) 465.
[33]  G. I. Bell, *Math. Biosc.* **16** (1973) 291.
[34]  L. Onsager and S. Machlup, *Phys. Rev.* **91** (1953) 1505.
[35]  D. Falkoff, *Ann. Physics* **4** (1958) 325.
[36]  B. Kursunoglu, *Am. J. Phys.* **34** (1966) 1043.
[37]  N.G. de Bruijn, *Asymptotic Methods in Analysis* (North-Holland, Amsterdam, 1961).
[38]  J.B. Jalickee, F.W. Wiegel and D.J. Vezzetti, *Phys. Fluids* **14** (1971) 1041.
[39]  F.W. Wiegel, *J. Stat. Phys.* **7** (1973) 213.
[40]  F.W. Wiegel, Ph.D. Dissertation, University of Amsterdam (1973), unpublished.
[41]  N.G. van Kampen, *J. Stat. Phys.* **17** (1977) 71.
[42]  H. Dekker and N.G. van Kampen, *Phys. Lett.* **73A** (1979) 374.
[43]  B. Caroli, C. Caroli and B. Roulet, *J. Stat. Phys.* **21** (1979) 415.

# III. MACROMOLECULES

Macromolecules are very long, chain-like molecules, which are the basic material of which living beings consist. These molecules are themselves like an incarnation of the concepts of path integration. It is especially the statistical physics of macromolecules which has benefited from the application of path-integral methods, and some of the outstanding problems in this branch of physical chemistry have been solved with such techniques. The aim of this chapter is to make the reader familiar with these problems and methods of solution to such an extent that he can start doing calculations of his own.

## 3.1. The free random walk model

Macromolecules are strings of small groups of atoms, which are called monomers, or repeating units. The successive repeating units of a macromolecule are strongly bound by chemical bonds, hence their mutual distance is almost constant. However, in many cases the repeating units can easily be rotated with respect to each other, so that the molecule as a whole—although of a fixed length—is extremely flexible. These two features of real macromolecules form the basis of the random walk model of macromolecular statistics, in which a configuration of a macromolecule consisting of $N$ identical repeating units is represented by a random walk of $N$ steps, each of a length $l$ which equals the average distance between two successive repeating units in the molecule. Depending on the interactions between the repeating units one should impose certain constraints on the random walk configurations. In a first approximation such monomer-monomer interaction effects are neglected altogether and, hence, the random walks considered are free.

Denote the probability density that a free random walk which starts at the origin of coordinates will reach the point $\mathbf{r}$ after $N$ steps by $P(\mathbf{r}, N)$. The first of these probability distributions is

$$P(\mathbf{r}, 1) = (4\pi l^2)^{-1} \delta(|\mathbf{r}| - l) , \tag{1.1a}$$

where $\delta(x)$ denotes the one-dimensional Dirac delta function with the properties

$$\delta(x) = 0 , \qquad (x \neq 0) , \tag{1.1b}$$

$$\delta(x) = +\infty , \qquad (x = 0) , \tag{1.1c}$$

$$\int_{-\infty}^{+\infty} \delta(x) \, dx = 1 . \tag{1.1d}$$

For $N > 1$ the distributions can be calculated in a recursive way from the relation

$$P(\mathbf{r}, N + 1) = (4\pi l^2)^{-1} \int P(\mathbf{r}', N) \delta(|\mathbf{r} - \mathbf{r}'| - l) \, d^3r' , \tag{1.2}$$

which simply expresses the fact that the probability $P(\mathbf{r}, N+1) \, d^3r$ that step $N+1$ ends in a vicinity $d^3r$ of the point $\mathbf{r}$ equals the probability $P(\mathbf{r}', N) d^3r'$ that step $N$ ends in a vicinity $d^3r'$ of $\mathbf{r}'$, times the transition probability $(4\pi l^2)^{-1} \delta(|\mathbf{r} - \mathbf{r}'| - l) \, d^3r$ that step $N + 1$ ends in $d^3r$, integrated over all volume elements $d^3r'$. The actual calculation of the functions $P(\mathbf{r}, N)$ can now proceed in various ways.

(a) *Exact calculation*: The Fourier transform $\tilde{P}(\mathbf{k}, 1)$ of $P(\mathbf{r}, 1)$ can be calculated directly from (1.1)

$$\tilde{P}(\mathbf{k}, 1) \equiv \int P(\mathbf{r}, 1) \exp(i\mathbf{k} \cdot \mathbf{r}) \, d^3r = \frac{\sin kl}{kl} , \tag{1.3}$$

where $k$ denotes the length of the vector $\mathbf{k}$. As $P(\mathbf{r}, N)$ is related to $P(\mathbf{r}, 1)$ by an $(N - 1)$ fold convolution product, its Fourier transform $\tilde{P}(\mathbf{k}, N)$ is the $N$th power of $\bar{P}(\mathbf{k}, 1)$. Hence one finds

$$\bar{P}(\mathbf{k}, N) = \left[\frac{\sin kl}{kl}\right]^N , \tag{1.4}$$

which can be inverted to give

$$P(\mathbf{r}, N) = (2\pi)^{-3} \int \left(\frac{\sin kl}{kl}\right)^N \exp(-i\mathbf{k} \cdot \mathbf{r}) \, d^3k \tag{1.5}$$

as an exact integral representation for the probability distribution.

(b) *Asymptotic behavior*: The asymptotic behavior of the probability distribution for $N \gg 1$ follows from the behavior of the Fourier transform for $kl \ll 1$. Expanding (1.3) for small $kl$ one finds

$$\bar{P}(\mathbf{k}, 1) = 1 - \frac{1}{6} k^2 l^2 + O(k^4 l^4) . \tag{1.6}$$

$$= \exp\left(-\frac{1}{6} k^2 l^2\right) + O(k^4 l^4) .$$

Substitution into (1.4) and (1.5) gives

$$\bar{P}(\mathbf{k}, N) \cong \exp\left(-\frac{1}{6} N k^2 l^2\right) , \qquad (N \gg 1) , \tag{1.7}$$

$$P(\mathbf{r}, N) \cong \left(\frac{2}{3} \pi N l^2\right)^{-3/2} \exp\left(-\frac{3r^2}{2Nl^2}\right) , \qquad (N \gg 1) . \tag{1.8}$$

This formula shows that one endpoint of a non-interacting macromolecule has a Gaussian distribution with respect to the position of the other endpoint.

(c) *The differential equation*: Another way to find the result (1.8) — and one that can be generalized to more complicated situations — consists of noting that for $N \gg 1$ the probability density $P(\mathbf{r}', N)$ in (1.2) will be a smooth function of $\mathbf{r}'$. Hence one can expand it in a Taylor series around $\mathbf{r}' = \mathbf{r}$

$$P(\mathbf{r}', N) = P(\mathbf{r}, N) + \sum_{i=1}^{3} (r_i' - r_i) \frac{\partial P}{\partial r_i} + \tag{1.9}$$

$$\frac{1}{2} \sum_{i,j=1}^{3} (r_i' - r_i)(r_j' - r_j) \frac{\partial^2 P}{\partial r_i \, \partial r_j} + \dots ,$$

$$P(\mathbf{r}, N + 1) = P(\mathbf{r}, N) + \partial P(\mathbf{r}, N)/\partial N + \dots , \tag{1.10}$$

where $r_i$ denotes the $i$th Cartesian coordinate of the vector $\mathbf{r}$ and where we treated $N$ as a continuous variable. Substitution into (1.2) gives

$$\frac{\partial P}{\partial N} \cong \frac{l^2}{6} \triangle P + O(l^4 \triangle^2 P) \ , \tag{1.11a}$$

where $\triangle$ denotes the Laplacian operator $\partial^2/\partial x^2 + \partial^2/\partial y^2 + \partial^2/\partial z^2$. Now note that $\partial P/\partial N$ and $l^2 \triangle P$ will be of order $N^{-1}$, but $l^4 \triangle^2 P$ is of order $N^{-2}$. Hence in the limit $N \to \infty$, $l \to 0$, $Nl^2 = $ constant, one has rigorously

$$\frac{\partial P}{\partial N} = \frac{l^2}{6} \triangle P \ , \tag{1.11b}$$

which has the Gaussian distribution (1.8) as its fundamental solution, as we showed in Chapter I.

(d) *Path integral*: Comparison of (1.11b) with (I.2.3) shows that one can write the probability density of the endpoint of the chain molecule in the form of a Wiener path integral

$$P(\mathbf{r}, N) = \int_{\mathbf{0},0}^{\mathbf{r},N} \exp\left\{ - \int_0^N L_0 \, d\nu \right\} d[\mathbf{r}(\nu)] \ , \tag{1.12a}$$

$$L_0 = \frac{3}{2l^2} \left( \frac{d\mathbf{r}}{d\nu} \right)^2 \ . \tag{1.12b}$$

The integrand in the Wiener integral is a measure for the number density of polymer configurations in which the coarse-grained shape of the molecule is close to some continuous and differentiable curve $\mathbf{r}(\nu)$. Of course, if a finer description of the chain configurations is necessary one should not use either the differential equation (1.11b) or the path integral (1.12) but instead use the recursive relation (1.2). For further discussions of free random walks the reader should consult the review paper of Chandrasekhar [I-44], the reprint collection in Wax [I-45] and the monograph by Barber and Ninham [I-46].

### 3.2 Dust, adsorption, and the boundary conditions for polymer statistics

Consider now the case in which space is filled with "dust," i.e., small impenetrable particles of any size which act as an excluded volume for the macromolecule. Let $f(\mathbf{r})$ denote the fraction of space from which the endpoints of the successive repeating units of the molecule are excluded. In this case the recursive relation (1.2) for free polymers has to be replaced by

$$P(\mathbf{r}, N + 1) = (4\pi l^2)^{-1} \{1 - f(\mathbf{r})\} \int P(\mathbf{r}', N) \, \delta(|\, \mathbf{r} - \mathbf{r}'\,| - l) \, d^3r'$$

$$(2.1)$$

for polymers in such a "dust-filled" space. Note that $P(\mathbf{r}, N) \, d^3r$ is now defined as the fraction of free random walk configurations that end in $d^3r$ and that do not intersect dust anywhere. Following the procedure of Section 3.1c one finds the differential equation

$$\frac{\partial P}{\partial N} = -fP + \frac{l^2}{6} (1 - f) \triangle P + (1 - f) \, O(l^4 \triangle^2 P) \ . \qquad (2.2)$$

Hence in the limit

$$N \to \infty \ , \quad l \to 0 \ , \quad f \to 0 \ , \qquad (2.3a)$$

$$Nl^2 = \text{constant} \ , \qquad (2.3b)$$

$$f/l^2 = \text{constant} \ , \qquad (2.3c)$$

the differential equation becomes

$$\frac{\partial P}{\partial N} = \frac{l^2}{6} \triangle P - fP \ , \qquad (2.4)$$

and the path integral is

$$P(\mathbf{r}, N) = \int_{\mathbf{0},0}^{\mathbf{r}, N} \exp(- \int_0^N L_f \, d\nu) \, d[\mathbf{r}(\nu)] \ , \qquad (2.5a)$$

$$L_f = \frac{3}{2l^2} \left(\frac{d\mathbf{r}}{d\nu}\right)^2 + f(\mathbf{r}) \ . \qquad (2.5b)$$

It is important at this point to stress a subtle difference between the statistics of Brownian motion paths, which formed the subject of Chapter I, and the statistics of macromolecules in a space which is partly filled with impenetrable matter. This difference shows up in the boundary conditions that have to be imposed on the solutions of the differential equations. Consider the density of Brownian particles near a hard wall. Here the boundary condition should express the fact that the particle current density $- D \nabla c$ should have no component perpendicular to the wall, so

$$\nabla c \parallel \text{hard wall} \qquad (\text{Brownian particles}). \qquad (2.6)$$

Next we consider the probability distribution $P(\mathbf{r}, N)$ of the endpoint of a polymer near a hard wall. This wall is simply modelled by taking in (2.4,5) the limit $f(\mathbf{r})/l^2 \to +\infty$ inside the wall, hence the path integral shows that

$$P = 0 \quad \text{at a hard wall,} \quad \text{(polymers).} \tag{2.7}$$

This is analogous to the boundary condition for Brownian particles at the boundary of a region of space in which the absorption probability per unit time is infinite

$$c = 0 \quad \text{at an absorbing wall,} \quad \text{(Brownian particles).} \tag{2.8}$$

To the author's knowledge the reflecting boundary condition (2.6) for Brownian particles does not have an equivalent in polymer statistics.

It should be clear from the discussion in this section how problems related to polymers in the presence of dust can be treated by solving the differential equation (2.4), using the boundary condition (2.7) if a hard wall is present. Now, in many cases of practical interest the polymer is in solution and the "wall" is some surface with a lot of structure of its own, for example a cell membrane. In these cases attractive forces exist, which favor the "condensation" of the molecule on the surface. This condensation is counteracted by the heat motion which favors those configurations that correspond to large volumes in phase space, i.e., configurations in which the molecule is in solution, away from the membrane. The resulting adsorption-desorption phase transition is the outcome of these two competing processes. A general discussion of adsorption can be found in the monograph of de Gennes [1]. Rubin has given rigorous solutions to certain lattice problems related to the adsorption of a polymer to a plane, a line or a point [2]. The essential features of these rigorous solutions are easily recovered using the method of path integration in the following way. Let the potential of the attractive force per monomer be denoted by $V(\mathbf{r})$. For a configuration of the polymer which is characterized by the positions $\mathbf{r}_0$, $\mathbf{r}_1, \ldots \mathbf{r}_N$ of the $N + 1$ endpoints of the $N$ monomers, the Boltzmann factor can be written in the continous form

$$\exp\left\{-\beta \sum_{j=0}^{N} V(\mathbf{r}_j)\right\} \cong \exp\left\{-\beta \int_0^N V(\mathbf{r}(\nu)) \, d\nu\right\}, \tag{2.9}$$

where $\beta = (k_B T)^{-1}$ with $k_B$ denoting Boltzmann's constant and $T$ the absolute temperature. Hence, using (1.12) the configuration sum $Q_V$ for the polymer with fixed endpoints is given by the Wiener integral

$$Q_V(\mathbf{r}_N, N | \mathbf{r}_0, 0) = \int_{\mathbf{r}_0, 0}^{\mathbf{r}_N, N} \exp\left\{-\frac{3}{2l^2} \int_0^N \left(\frac{d\mathbf{r}}{d\nu}\right)^2 d\nu - \beta \int_0^N V(\mathbf{r}(\nu)) d\nu\right\} d[\mathbf{r}(\nu)] .$$
$$\tag{2.10}$$

Using the results of Section 1.4 we see that $Q_V$ can also be interpreted as the propagator of the partial differential equation

$$\frac{\partial c}{\partial N} = \frac{l^2}{6} \Delta c - \beta V(\mathbf{r})c . \qquad (2.11)$$

Of course, using the continuous form of the Boltzmann factor (2.9)— which leads to a differential equation of the diffusion type—essentially implies a limiting process similar to (2.3)

$$N \to \infty, \quad l \to 0 , \quad \beta V \to 0, \qquad (2.12a)$$

$$Nl^2 \quad = \text{constant}, \qquad (2.12b)$$

$$\beta V/l^2 = \text{constant}. \qquad (2.12c)$$

If $\beta V$ is not of order $l^2$ one has to use an integral equation similar to (2.1). Now, taking (2.11) for granted, we see that its propagator has the bilinear expansion

$$Q_V(\mathbf{r}_N, N \mid \mathbf{r}_0, 0) = \sum_n f_n(\mathbf{r}_N) f_n^*(\mathbf{r}_0) \exp(-E_n N) \qquad (2.13)$$

in terms of the orthonormal eigenfunctions and eigenvalues of the equation

$$\left[ -\frac{l^2}{6} \Delta + \beta V(\mathbf{r}) \right] f_n(\mathbf{r}) = E_n f_n(\mathbf{r}) , \qquad (2.14a)$$

with the boundary condition (2.7) at all hard walls

$$f_n = 0 \quad \text{at a hard wall}. \qquad (2.14b)$$

The solution of the absorption problem is thus reduced to the solution of the eigenfunction problem (2.13).

In the case of adsorption to a plane surface the $x$ coordinate can be taken to be perpendicular to the surface. Let the surface be situated at $x=0$; this implies $f_n(0, y, z) = 0$. The interaction energy is now a function $V(x)$ of $x$ only, and after separation of the variables the essential part of the eigenvalue problem is the $x$-dependent part

$$\left[ -\frac{l^2}{6} \frac{d^2}{dx^2} + \frac{V(x)}{k_B T} \right] F_m(x) = \lambda_m F_m(x) , \qquad (2.15)$$

$$F_m(0) = 0 . \qquad (2.16)$$

For an attractive force $V(x)$ will qualitatively have the form of a potential well near the membrane, and $V(x) \to 0$ if $x \to \infty$. If one lets $T$ decrease from $+\infty$ to $0$ one finds two regimes in each of which the adsorption phenomenon has different characteristics:

(1)   For $T > T_c$, the potential well in $\beta V$ has no bound states. Thus all the $\lambda_m$ are positive and the corresponding $F_m$ have the asymptotic form

$$F_m(x) \cong A_m \sin(\sqrt{6\lambda_m/l^2}\, x) + B_m \cos(\sqrt{6\lambda_m/l^2}\, x)\,, \quad (x \to \infty)\,. \quad (2.17)$$

This means physically that the configurations of the macromolecule are very open and not localized near the surface, i.e., the polymer is desorbed from the surface.

(2)   For $T < T_c$ the potential well in $\beta V$ has one or more bound states with eigenvalues $\lambda_m < 0$. For $N \gg 1$ the sum in (2.13) will be dominated by the ground state eigenfunction $F_0(x)$

$$Q_V(\mathbf{r}_N, N \mid \mathbf{r}_0, 0) = \text{(function of the } y \text{ and } z \text{ coordinates)}$$

$$\times\, F_0(x_N)\, F_0^*(x_0)\, \exp(-\lambda_0 N)\,. \quad (2.18)$$

Moreover, the asymptotic behavior of the bound state(s) is of the form

$$F_0(x) \cong A_0 \exp\left(- \sqrt{6|\lambda_0|/l^2}\, x\right)\,, \quad (x \to \infty)\,. \quad (2.19)$$

As this function is "localized" in the potential well the configurations of the polymer are now localized in a narrow layer near the adsorbing surface. The thickness of this layer is of order

$$\text{(thickness)} \cong \frac{l}{\sqrt{6|\lambda_0|}}\,. \quad (2.20)$$

It should be clear that an enormous variety of adsorption problems (different geometries, different interaction potentials $V(\mathbf{r})$) can be solved explicitly with the method embodied in Eqs. (2.13–14). A fairly realistic case is that in which the polymer carries a fixed charge $q$ per monomer, and where the adsorbing membrane carries fixed charges with a constant surface density $\sigma$. Due to screening effects in the solvent the electrostatic interaction energy between a monomer and a surface element of area $d^2S$ equals

$$d^2V = \frac{q\sigma}{\varepsilon r} \exp(-\kappa r)\, d^2S\,, \quad (2.21)$$

where $\varepsilon$ is the dielectric constant of the solvent and $\kappa^{-1}$ the Debye screening length. This case can be solved exactly [3]; the adsorption-desorption phase transition is found to occur when $\sigma q < 0$ at a temperature

$$k_B T_c = \frac{48\pi\, |\sigma q|}{j_{0.1}^2\, \kappa^3 l^2 \varepsilon}\,, \quad (2.22)$$

where $j_{0,1} \simeq 2.4048$ is the first positive zero of the Bessel function $J_0(\xi)$. We shall not pursue the details of this calculation here as it leads away from path integration and into the physics of the cell membrane [4].

## 3.3 The excluded volume problem

### 3.3.1. *The method of the most probable configuration*

When the forces between the repeating units of a macromolecule are taken into account the calculation of the number of chain configurations becomes much more difficult. The most fundamental problem of this type is the excluded volume problem which in its simplest form can be formulated as follows. Consider all random walk chain configurations of $N$ steps, in a $d$-dimensional space, which start at the origin of coordinates. Around each of the $N$ endpoints of the monomers we imagine a small hard sphere of radius $a/2 < l/2$. Hence, around each endpoint there is a volume $\gamma_d$ which is forbidden for all other endpoints. Here $\gamma_2 = \pi a^2$, $\gamma_3 = 4/3\pi a^3$, etc. The excluded volume problem consists of calculating the total number of non-self intersecting chain configurations, i.e., the fraction of random walk configurations which are self-avoiding.

This problem has not been solved exactly for any dimension larger than one (for $d = 1$ the solution is trivial). There are, however, several methods for an approximate solution available, which we shall briefly discuss in the following pages. When cast in the language of path integration the appropriate approximation techniques are simple, transparent and give quite accurate results.

The simplest approximation method, which we shall call the method of the most probable configuration, is due to de Gennes [5], also cf. Appendix E of Wiegel [6]. This method relies on two simple and fairly plausible assumptions: (a) The statistical properties of polymers with an excluded volume are approximately the same as those of polymers without an excluded volume in a space in which a volume fraction $f(\mathbf{r})$ is occupied by small hard spheres of volume $\gamma_d$, *provided $f(\mathbf{r})$ is calculated from the distribution of repeating units in a self-consistent way.* (b) These statistical properties can approximately be calculated from the most probable chain configuration.

According to (2.5) the most probable configuration is that curve $\mathbf{r}^*(\nu)$ that minimizes the integral

$$S[\mathbf{r}(\nu)] \equiv \int_0^N \left\{ \frac{1}{4D_d} \left( \frac{d\mathbf{r}}{d\nu} \right)^2 + f(\mathbf{r}) \right\} d\nu \ , \tag{3.1}$$

(where $D_2 = l^2/4, D_3 = l^2/6$, etc.) under the boundary condition $\mathbf{r}(0) = 0$. Note that there is no boundary condition at $\nu = N$ as the endpoint $\mathbf{r}(N)$ of the molecule can be anywhere in space. The minimization of the integral (3.1) is the standard problem of the calculus of variations, cf. [II–25], where it is shown that the minimum should be a solution of the Euler-Lagrange equation

$$(2D_d)^{-1} \frac{d^2\mathbf{r}^*}{d\nu^2} = \nabla f(\mathbf{r}^*) , \qquad (3.2)$$

the gradient being taken with respect to $\mathbf{r}^*$. This equation is formally equivalent to the equation of motion of a classical particle with mass $(2D_d)^{-1}$, coordinate $\mathbf{r}^*$, moving in time $\nu$ in a scalar external potential

$$V(\mathbf{r}^*) = -f(\mathbf{r}^*) . \qquad (3.3)$$

The total energy $E$ will be a constant of the motion, hence

$$(4D_d)^{-1} \left(\frac{d\mathbf{r}^*}{d\nu}\right)^2 = E + f(\mathbf{r}^*) . \qquad (3.4)$$

For reasons of symmetry the motion of the "particle" will be along a radius vector in the direction of the initial "velocity" $(d\mathbf{r}^*/d\nu)_{\nu=0}$.

The function $f$ can now be determined in a self-consistent way as follows. Consider a spherical shell around the origin with radius $r$, thickness $dr$ and area $A_d r^{d-1}$ (where $A_2 = 2\pi$, $A_3 = 4\pi$, etc.). The average number of monomers in this shell equals $(A_d/\gamma_d)r^{d-1} f(r)dr$. On the other hand, the average number of monomers equals $d\nu = (dr/d\nu)^{-1}dr$, where $r \equiv r^*(\nu)$, the most probable configuration. Hence one finds

$$f(r) = \frac{\gamma_d}{A_d} r^{1-d} \left(\frac{dr}{d\nu}\right)^{-1} , \qquad (3.5)$$

and the integral (3.1) can also be written in the form

$$S[\mathbf{r}(\nu)] = \int_0^N \left\{ \frac{1}{4D_d} \left(\frac{dr}{d\nu}\right)^2 + \frac{\gamma_d}{A_d} r^{1-d} \left(\frac{dr}{d\nu}\right)^{-1} \right\} d\nu . \qquad (3.6)$$

We have to find the minimum of this integral for all monotonically increasing functions $r(\nu)$ with $r(0) = 0$. It is straightforward to verify that the Euler-Lagrange equation of the variational problem has the form

$$\left\{ \frac{1}{2D_d} + \frac{2\gamma_d}{A_d} r^{1-d} \left(\frac{dr}{d\nu}\right)^{-3} \right\} \frac{d^2r}{d\nu^2} + 2(d-1) \frac{\gamma_d}{A_d} r^{-d} \left(\frac{dr}{d\nu}\right)^{-1} = 0 . \qquad (3.7)$$

This rather formidable-looking equation has a simple solution of the form

$$r(\nu) = K\nu^\alpha ,\qquad(3.8)$$

where $K$ and $\alpha$ are positive constants. Substitution gives immediately

$$\alpha = \frac{3}{d + 2} ,\qquad(3.9)$$

$$K = \left(\frac{A_d}{8\gamma_d D_d}\right)^{-1/(d+2)} \left(\frac{3}{d+2}\right)^{-3/(d+2)} .\qquad(3.10)$$

Hence we have found that the two endpoints of a macromolecule with excluded volume have a linear distance $R$ the square of which grows with $N$ like

$$R^2 \cong K^2 N^{6/(d+2)} .\qquad(3.11)$$

The prediction of a "critical exponent" 3/2 for $d = 2$ and 6/5 for $d = 3$ is in satisfactory agreement with computer enumerations of non-self intersecting chain configurations (cf. [7, 8] and the papers quoted therein). It is certainly remarkable how easy it is to derive the value of this critical exponent from the path integral with the method of the most probable configuration.

### 3.3.2. *Equivalence with a random potential problem*

The approximation method just discussed, although quite elegant, has the disadvantage that it is not easy to see how it might be improved in a systematic way. For this reason a more systematic approach is needed. In this subsection we dicuss a method which "translates" the excluded volume problem in a random potential problem. This approach has been pioneered especially by Edwards [9, 10], and somewhat later by Freed [11], and has given rise to a huge literature.

Essentially the idea is to represent the self interaction of the polymer by a sum of pair interactions $V(\mathbf{r}_i - \mathbf{r}_j)$ between each pair $i, j$ of monomers located at $\mathbf{r}_i, \mathbf{r}_j$. We shall often take

$$V(\mathbf{r}_i - \mathbf{r}_j) = +V_0 \quad \text{if} \quad |\mathbf{r}_i - \mathbf{r}_j| \le a < l ,$$

$$= 0 \quad \text{otherwise},\qquad(3.12)$$

where $V_0$ is a large positive constant, but the formalism is general. The Boltzmann factor for the configuration $\mathbf{r}_0, \mathbf{r}_1, \ldots, \mathbf{r}_N$ is

$$\exp\left\{-\beta \sum_{1 \le i < j \le N} V(\mathbf{r}_i - \mathbf{r}_j)\right\} .$$

Now we introduce Gaussian random functions $\phi(\mathbf{r})$ with zero average and with a covariance

$$<\phi(\mathbf{r}) \, \phi(\mathbf{r}')> = -\beta V(\mathbf{r} - \mathbf{r}') , \qquad (3.13)$$

where $< >$ denotes the average over the space of these functions. The properties of these random functions will be discussed in full detail in Chapter VI. All we need is the identity (VI.2.4) which can be written as

$$\exp\left\{-\beta \sum_{1 \le i \le j \le N} V(\mathbf{r}_i - \mathbf{r}_j)\right\} = \exp(+\tfrac{1}{2} \beta N V_0) <\exp\left\{-\sum_{j=1}^{N} \phi(\mathbf{r}_j)\right\}> .$$
$$(3.14)$$

Now the configuration sum for a polymer with an excluded volume, which we shall denote by $\Pi(\mathbf{r}, N)$, is according to (1.12) given by the path integral over the Boltzmann factor

$$\Pi(\mathbf{r}, N) = \exp(+\frac{1}{2} \beta N V_0) \qquad (3.15)$$

$$\times \int_{\mathbf{0},0}^{\mathbf{r},N} \exp\left\{-\frac{1}{4D_d} \int_0^N \left(\frac{d\mathbf{r}}{d\nu}\right)^2 d\nu\right\} <\exp\left\{-\sum_{j=1}^{N} \phi(\mathbf{r}_j)\right\}> d[\mathbf{r}(\nu)] .$$

Just as in (2.9) we shall write the exponential in the continuous notation

$$\sum_j \phi(\mathbf{r}_j) \cong \int_0^N \phi(\mathbf{r}(\nu)) \, d\nu . \qquad (3.16)$$

Noting that the average over the $\phi$ can be taken after the path integral over the $\mathbf{r}(\nu)$ has been performed, one finds

$$\Pi(\mathbf{r}, N) = \exp(+\frac{1}{2} \beta N V_0) < G_\phi(\mathbf{r}, N) > , \qquad (3.17)$$

where

$$G_\phi(\mathbf{r}, N) \equiv \int_{\mathbf{0},0}^{\mathbf{r},N} \exp\left\{-\frac{1}{4D_d} \int_0^N \left(\frac{d\mathbf{r}}{d\nu}\right)^2 d\nu - \int_0^N \phi(\mathbf{r}(\nu)) \, d\nu\right\} d[\mathbf{r}(\nu)] . \quad (3.18)$$

But this path integral is the propagator of a diffusion equation with an annihilation term; actually one has, from Section 1.4,

$$\left[\frac{\partial}{\partial N} - D_d \Delta + \phi(\mathbf{r})\right] G_\phi(\mathbf{r}, N) = \delta(\mathbf{r}) \, \delta(N) . \qquad (3.19)$$

As the annihilation term $\phi(\mathbf{r})$ is itself a random function (a "random potential" in the sloppy jargon of theoretical physics) the excluded volume problem has been translated into a random potential problem.

It should be clear that the contents of this subsection are rather formal, and that the main result, Eq. [3.17], only shows the equivalence of two unsolved problems: the excluded volume problem and the problem of calculating the average propagator in a random potential. The only advantage of such an equivalence is that approximation techniques developed for one problem can be of use to the student of the other problem. We shall not puruse these approximations in any detail, but merely make several comments which might guide the reader to find his way through the literature (which is actually quite bewildering):

(1)   First of all it should be pointed out that, as $\beta V$ will be a predominantly repulsive potential, the random function $\phi(\mathbf{r})$ will take complex values, with large imaginary parts (cf. Section 6.1). Hence the "random potential" is not a very physical entity.

(2)   Secondly, after $G_\phi(\mathbf{r}, N)$ has been calculated explicitly one still has to perform the average over $\phi$. This can be done using Eq. (VI.1.14) and some kind of saddle-point method. It turns out that the field $\Phi(\mathbf{r})$ which is the main saddle-point actually has real values only [12].

(3)   In order to make some progress towards an explicit result Edwards [9, 10] assumed that only the main saddle-point $\Phi$ contributes, and, moreover, that it has spherical symmetry around the initial point of the polymer chain. The first approximation implies that the contributions of all secondary saddle points, which played such an important role in Section 2.4, are neglected. The second approximation is inspired by the idea that $\Phi$ has something to do with the average volume fraction $f(r)$ in Subsection 3.3.1, which approximately had spherical symmetry too.

(4)   Freed [11] has clarified the significance of some of these approximations. Since the early seventies the literature on the path-integral approach to be excluded volume problem has grown tremendously and could easily form the subject of a separate monograph. Of the very recent literature we merely quote an important paper by Muthukumar and Nickel [13].

### 3.3.3   *Diagram expansions and path integrals*

In many problems one can use a path integral to sum a diagram series in a formal way. Conversely, a path integral can often be expanded in some type of perturbation series, the terms of which can be represented by diagrams; an example is the series in Eq. (I.5.4), the terms of which can be represented by Feynman diagrams. We shall use the excluded volume problem to demonstrate this technique.

The starting point is the following expression for the configuration sum of a polymer the monomers of which interact through the pair potential $V(\mathbf{r}_i - \mathbf{r}_j)$

$$\Pi(\mathbf{r}, N) = \int d^3r_1 \int d^3r_2 \ldots \int d^3r_{N-1} \exp\left\{-\beta \sum_{i<j} V(\mathbf{r}_i - \mathbf{r}_j)\right\}$$

$$\times \prod_{i=0}^{N-1} (4\pi l^2)^{-1} \delta(|\mathbf{r}_{i+1} - \mathbf{r}_i| - l) \ . \tag{3.20}$$

This is essentially the same expression which was the subject of the previous subsection, apart from a more detailed description of the random walk model of the chain, the continuous limit of which will be taken somewhat later in the calculation.

Now, if the potential $V$ has a hard core it is more convenient to try a perturbation expansion in terms of the Mayer function

$$f(\mathbf{r}) = \exp\{-\beta V(\mathbf{r})\} - 1, \tag{3.21}$$

which has the value $-1$ for $V = +\infty$. Of course, for $\beta V \ll 1$ one finds $f(\mathbf{r}) \simeq -\beta V(\mathbf{r})$ and the results of this subsection become identical to those of Subsection 3.3.2.

Substituting (3.21) into (3.20) and denoting the operation

$$\int d^3r_1 \int d^3r_2 \ldots \int d^3r_{N-1} \prod_{i=0}^{N-1} (4\pi l^2) \delta(|\mathbf{r}_{i+1} - \mathbf{r}_i| - l)$$

by the symbol [ ], one obtains the Mayer cluster expansion in the form

$$\Pi(\mathbf{r}, N) = \left[ \prod_{i<j} (1 + f(\mathbf{r}_i - \mathbf{r}_j)) \right]$$

$$= [1] + \sum_{i<j} [f(\mathbf{r}_i - \mathbf{r}_j)]$$

$$+ \sum_{i<j} \sum_{i'<j'} [f(\mathbf{r}_i - \mathbf{r}_j) f(\mathbf{r}_{i'} - \mathbf{r}_{j'})] + \ldots \tag{3.22}$$

The terms of this expansion can be represented by diagrams as follows.

Obviously, the process indicated by [ ] becomes a Wiener integral in the continuous limit. Hence $[1] = G_0(\mathbf{r}_N, N \mid \mathbf{0}, 0)$ as given by (1.8). This term will be represented by a directed solid line from the vertex $(\mathbf{0}, 0)$ to the vertex $(\mathbf{r}_N, N)$. The second term in (3.22) equals

$$[f(\mathbf{r}_i - \mathbf{r}_j)] = \int d^3r_i \int d^3r_j \, G_0(\mathbf{r}_N, N \mid \mathbf{r}_j, j) \, G_0(\mathbf{r}_j, j \mid \mathbf{r}_i, i)$$

$$\times f(\mathbf{r}_i - \mathbf{r}_j) \, G_0(\mathbf{r}_i, i \mid \mathbf{0}, 0), \tag{3.23}$$

and can be represented by the diagram in Fig. 3.1 in which the wavy line represents the Mayer function. The third term $[f(\mathbf{r}_i - \mathbf{r}_j) f(\mathbf{r}_{i'} - \mathbf{r}_{j'})]$ in the cluster expansion leads to three diagrams, depending on the relative order of the labels $i, j, i', j'$; they are indicated in Fig. 3.2.

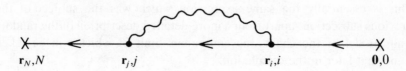

Fig. 3.1.  Diagram representing eq. (3.23). Solid lines represent the factors $G_0$, wavy lines represent factors $f$.

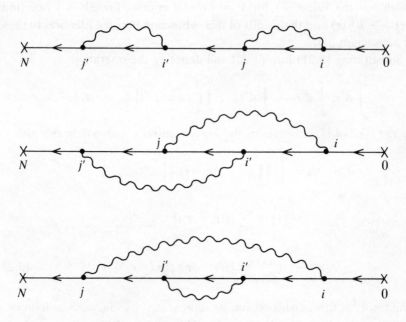

Fig. 3.2.  Three diagrams which together are equal to $[f(\mathbf{r}_i - \mathbf{r}_j) f(\mathbf{r}_{i'} - \mathbf{r}_{j'})]$.

From the examples the following rules for the diagrams are apparent:
(1)  Take $0, 1, 2, \ldots$ interaction lines, indicated by wavy lines and numerically equal to the Mayer function $f(\mathbf{r}_1 - \mathbf{r}_2)$.
(2)  Attach their endpoints to one directed solid line in all possible ways. The endpoints are dots.
(3)  A solid line stands for a free propagator $G_0(\mathbf{r}_2, i_2 \mid \mathbf{r}_1, i_1)$. We shall

take this to vanish if $i_2 \le i_1$. By doing this we make an error in the configuration sum by omitting the contributions of all configurations in which a cluster of three or more monomers are simultaneously in interaction (connected by wavy lines). These configurations will be unimportant for a polymer unless $V(\mathbf{r})$ is sufficiently strongly attractive to force the polymer to collapse into a globule.

(4) Sum all indices $i$ from 1 to $\infty$ and integrate all endpoints over all space.

Now let $\phi(\mathbf{r})$ be a Gaussian random function with vanishing average and covariance

$$<\phi(\mathbf{r}) \, \phi(\mathbf{r}')> = f(\mathbf{r} - \mathbf{r}').  \tag{3.24}$$

Their properties are discussed in Chapter VI. Using their decomposition property VI.1.4 we find that the diagram in Fig. 3.1 equals

$$\int d^3 r_j \int d^3 r_i \int dj \int di <G_0(\mathbf{r}, N \mid \mathbf{r}_j, j) \, \phi(\mathbf{r}_j) \, G_0(\mathbf{r}_j, j \mid \mathbf{r}_i, i)$$

$$\times \, \phi(\mathbf{r}_i) \, G_0(\mathbf{r}_i, i \mid \mathbf{0}, 0)> ,  \tag{3.25}$$

the sum of the diagrams in Fig. 3.2 equals

$$\int d^3 r_l \int d^3 r_k \int d^3 r_j \int d^3 r_i \int dl \int dk \int dj \int di$$

$$\times < G_0(\mathbf{r}, N \mid \mathbf{r}_l, l) \, \phi(\mathbf{r}_l) \, G_0(\mathbf{r}_l, l \mid \mathbf{r}_k, k) \, \phi(\mathbf{r}_k) \, G_0(\mathbf{r}_k, k \mid \mathbf{r}_j, j)$$

$$\times \, \phi(\mathbf{r}_j) \, G_0(\mathbf{r}_j, j \mid \mathbf{r}_i, i) \, \phi(\mathbf{r}_i) \, G_0(\mathbf{r}_i, i \mid \mathbf{0}, 0) > ,  \tag{3.26}$$

and so on to all orders of perturbation theory! The dots have now become the factors $(-\phi)$, so in a way one can say that the dots now "interact" with the random field independently from another. In this way we have found that all original diagrams are recovered if one sums over all the terms of the form (3.25, 26) with an even number of dots. However, all such terms with an odd number of dots may be added, as the average of a product of an odd number of Gaussian random functions vanishes [cf. Eq. (VI.1.4)].

Summarizing our findings up till now we have the identity

$$\Pi(\mathbf{r}, N) = < G_\phi(\mathbf{r}, N) >  \tag{3.27}$$

where $G_\phi(\mathbf{r}, N)$ is given by a series which can symbolically be written in the form

$$G_\phi = G_0 + \int G_0 \phi G_0 + \int \int G_0 \phi G_0 \phi G_0 + \dots  \tag{3.28}$$

But this implies that $G$ obeys the integral equation

$$G_\phi(\mathbf{r}, N) = G_0(\mathbf{r}, N \mid \mathbf{0}, 0) - \int dn' \int d^3r' \, G_0(\mathbf{r}, N \mid \mathbf{r}', n')$$

$$\times \, \phi(\mathbf{r}') \, G_\phi(\mathbf{r}', n' \mid \mathbf{0}, 0) \, . \tag{3.29}$$

Comparison with Section 1.4 shows that $G_\phi(\mathbf{r}, N)$ is the solution of the partial differential equation

$$\left[ \frac{\partial}{\partial N} - \frac{l^2}{6} \, \Delta + \phi(\mathbf{r}) \right] G_\phi(\mathbf{r}, N) = \delta(\mathbf{r}) \, \delta(N) \, . \tag{3.30}$$

The result (3.27) is, of course, essentially identical to our previous result (3.17). The derivation in this subsection shows that for realistic potentials $V$ the Gaussian random functions should not have covariance $-\beta V$ as in (3.13), but should have a covariance $\exp(-\beta V) - 1$. As we have remarked already, in the limit $\beta V \ll 1$ the difference between these two ways of "parametrization" of the intra-chain monomer-monomer interaction becomes negligible. In this limit the extra factor $\exp(+\frac{1}{2}\beta N V_0)$ in (3.17) also tends to unity.

In summary it should be stressed that the equivalence

$$\text{(diagram series)} \rightleftarrows \text{(path integral)} \tag{3.31}$$

which formed the subject of this subsection, can be used in two ways. This is also reflected in the history of the subject. Originally, around 1950, Feynman had developed path integrals which he expanded into series of diagrams ($\leftarrow$). The whole decade from 1955 to 1965 was infested by numerous attempts by many authors to sum diagram series directly. As the majority of these attempts failed, the period after 1965 showed a general trend to represent a diagrams series by a path integral ($\rightarrow$) and to try to evaluate the path integral by other means. "Other means" after 1970 especially indicates the method of the renormalization group, which we shall discuss in Chapter IX. It is somewhat amusing to note another reversal in the very recent literature, when people again try to expand path integrals in series of diagrams ($\leftarrow$)!

## 3.4    Steric repulsion between cells

When two cells (or colloidal particles) are brought in close apposition a force of steric repulsion arises, which is caused by macromolecules that are attached to one of the cell membranes. In this section we calculate the

force of steric repulsion per unit area with the method of the previous
section. The basic geometry, which is illustrated in Fig. 3.3, consists of two

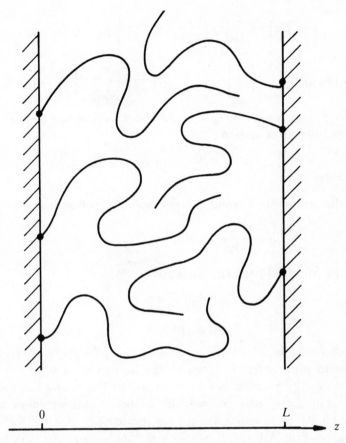

Fig. 3.3.   Basic geometry of steric repulsion between two cells, located at $z < 0$ and $z > L$.

parallel planes, each of area $A$, which represent the two cell membranes
(or the surfaces of the colloidal particles) in the contact area. We draw a
Cartesian set of coordinates with the $x$, $y$ axis parallel to the surface of the
membranes. The membranes will be represented by hard walls located at
$z = 0$ and $z = L$. To each of the membranes in the contact area $M$ flexible
polymers are attached with one of their endpoints. Each polymer is a
flexible chain of $N$ repeating units of length $l$; the endpoint of a repeating
unit excludes a spherical volume $\gamma = \frac{4}{3}\pi a^3$ from occupation by any other
repeating unit.

Using Eqs. (2.5a, b) and the fact that the fraction of space blocked by polymeric material is now a function $f(z)$ of the $z$ coordinate only, one finds that the configuration sum of one of the polymers with ends fixed at the origin on the left membrane, and at $\mathbf{r} = (x, y, z)$ equals

$$P(\mathbf{r}, N) = \left(\frac{2}{3}\pi Nl^2\right)^{-1} \exp\left\{-\frac{3}{2Nl^2}(x^2 + y^2)\right\} F(z, N \mid 0, 0) , \qquad (4.1)$$

$$F(z, N \mid 0, 0) = \int_{0,0}^{z,N} \exp\left[-\int_0^N \left\{\frac{3}{2l^2}\left(\frac{dz}{d\nu}\right)^2 + f(z)\right\} d\nu\right] d[z(\nu)] . \quad (4.2)$$

Of course, this Wiener integral could also be calculated from (2.4) and from the bilinear expansion

$$F(z, N \mid 0, 0) = \sum_n \phi_n^*(0)\, \phi_n(z) \exp(-\lambda_n N) , \qquad (4.3)$$

where the orthonormal eigenfunctions should be solved from

$$\left[-\frac{l^2}{6}\frac{d^2}{dz^2} + f(z)\right] \phi_n(z) = \lambda_n \phi_n(z) \qquad (4.4)$$

under the hard wall boundary conditions (2.7)

$$\phi_n(0) = 0 , \qquad (4.5a)$$

$$\phi_n(L) = 0 . \qquad (4.5b)$$

In order to close the set of equations and solve for $f(z)$ and $P(\mathbf{r}, N)$ one still has to express $f(z)$ in terms of the configuration sums $P(\mathbf{r}, \nu)$ for $0 \le \nu \le N$. This problem has been studied by Dolan and Edwards [14], both analytically and numerically. These authors noticed that $F(z, N \mid 0, 0)$ is proportional to the probability density for the $z$-coordinate of the $N$th repeating unit if the chain starts at $z = 0$. We now pick one of the repeating units, and ask for the probability density $\rho(z, \nu)$ associated with its $z$-coordinate. From (4.2) one finds that

$$\rho(z, \nu) = \frac{\displaystyle\int_0^L F(z', N \mid z, \nu)\, F(z, \nu \mid 0, 0)\, dz'}{\displaystyle\int_0^L F(z', N \mid 0, 0)\, dz'} . \qquad (4.6)$$

Hence repeating unit $\nu$ contributes a term $\gamma\rho(z, \nu)/A$ to $f(z)$, and the total fraction of space blocked by the repeating units of this particular polymer equals

$$f_1(z) = \frac{\gamma}{A} \int_0^N \rho(z, \nu) \, d\nu \; . \tag{4.7}$$

Now, there are $M$ polymers attached to each membrane, hence

$$f(z) = \frac{M}{A} \gamma \int_0^N \rho(z, \nu) \, d\nu$$

+ a similar term due to the membrane on the right.    (4.8)

This in principle closes the set of equations.

In the rest of this section we shall pursue a simple approximation which leads to an analytical treatment of the problem. In order to introduce this approximation consider the eigenvalue problem (4.4, 5). The eigenvalues can be ordered in increasing order

$$\lambda_1 < \lambda_2 < \ldots \tag{4.9}$$

Their spacing can be estimated from the spacing of the "unperturbed" spectrum $\lambda_n^{(0)}$, defined by

$$-\frac{l^2}{6} \frac{d^2}{dz^2} \phi_n^{(0)} = \lambda_n^{(0)} \phi_n^{(0)} \; , \tag{4.10}$$

$$\phi_n^{(0)}(0) = \phi_n^{(0)}(L) = 0 \; . \tag{4.11}$$

The solution is

$$\phi_n^{(0)} = (2/L)^{1/2} \sin \frac{n\pi z}{L} \; , \qquad (n = 1, 2, 3, \ldots) \tag{4.12}$$

$$\lambda_n^{(0)} = \frac{n^2 \pi^2 l^2}{6L^2} \; . \tag{4.13}$$

Hence we estimate the distance between $\lambda_1$ and $\lambda_2$ by

$$\lambda_2 - \lambda_1 = O(l^2/L^2) \; . \tag{4.14}$$

A glance at (4.3) shows that for

$$Nl^2 \gg L^2 \tag{4.15}$$

the eigenfunction expansion can be approximated by its first term only

$$F(z, N \mid 0, 0) \cong \phi_1^*(0) \, \phi_1(z) \exp(-\lambda_1 N) \; . \tag{4.16}$$

As the same holds for most of the $F(z', N \mid z, \nu)$ which occur in (4.6) the density of the $\nu$th repeating unit is

$$\rho(z, \nu) \cong \frac{\displaystyle\int_0^L \phi_1^*(z)\,\phi_1(z')\,\phi_1^*(0)\,\phi_1(z)\,dz'}{\displaystyle\int_0^L \phi_1(z')\,\phi_1^*(0)\,dz'} = \phi_1^2(z) , \qquad (4.17)$$

where the orthonormality of the eigenfunctions was used, and the fact that the ground state is a real function. Upon substitution of (4.8) this gives

$$f(z) \cong \frac{2MN\gamma}{A}\,\phi_1^2(z) , \qquad (4.18)$$

the factor two being due to the fact that there are two membranes.

When this approximation is substituted into (4.4) one finds the non-linear eigenvalue problem

$$-\frac{l^2}{6}\frac{d^2\phi_1}{dz^2} + \frac{2MN\gamma}{A}\,\phi_1^3 = \lambda_1\phi_1 . \qquad (4.19)$$

This differential equation can be interpreted as the equation of motion of a particle with mass $l^2/6$, moving along the $\phi_1$ axis in a force field of magnitude

$$\text{force} = \frac{2MN\gamma}{A}\,\phi_1^3 - \lambda_1\phi_1 \qquad (4.20)$$

with scalar potential

$$V(\phi_1) = +\frac{1}{2}\lambda_1\phi_1^2 - \frac{MN\gamma}{2A}\,\phi_1^4 . \qquad (4.21)$$

This potential has a minimum at $\phi_1 = 0$ and two equal maxima at

$$\phi_1 = \pm\,\phi_{\max} = \pm\left(\frac{\lambda_1 A}{2MN\gamma}\right)^{1/2} . \qquad (4.22)$$

Combination with the boundary condition (4.5) shows that the ground state must have the qualitative form drawn in Fig. 3.4. Note that the fictitious particle should stay very close to the top of the hill in the potential at $\phi_m$ in order to satisfy the second boundary condition (4.5b). Hence we find that

$$\phi_1(z) \cong \left(\frac{\lambda_1 A}{2MN\gamma}\right)^{1/2} \qquad (4.23)$$

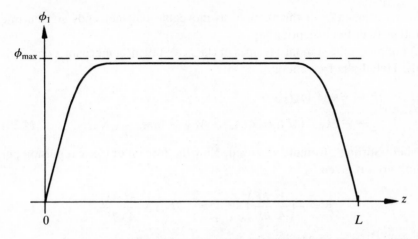

Fig. 3.4. Qualitative form of the ground state $\phi_1(z)$ of Eq. (4.4).

in almost the entire volume. The normalization condition

$$\int_0^L |\phi_1(z)|^2 \, dz = 1$$

now gives the lowest eigenvalue

$$\lambda_1 = \frac{2MN\gamma}{AL} \ . \tag{4.24}$$

When (4.16) is substituted into (4.1) and the resulting expression for $P(\mathbf{r}, N)$ integrated over $\mathbf{r}$, one finds that the configuration sum for one polymer with one end at the origin on the left membrane has the form

$$Q_1 = \int P(\mathbf{r}, N) \, d^3r \simeq C \exp(-\lambda_1 N) \ , \tag{4.25}$$

where $C$ is a constant. When the $2M$ initial endpoints of all $2M$ polymers are at fixed positions on one of the two membranes one thus finds for the configuration sum $Q_{2M}$ of the whole system

$$Q_{2M} = (M!)^{-2} C^{2M} \exp(-2\lambda_1 NM) \ . \tag{4.26}$$

On the other hand, when the $2M$ initial endpoints are not at fixed positions but free to move throughout the contact area $A$ the configuration sum is

$$Q_{2M} = (M!)^{-2} (CA/A_0)^{2M} \exp(-2\gamma_1 NM) \ , \tag{4.27}$$

where $A_0$ is a small area of the order of $a^2$. We shall use the last expression

in the remainder of this section, as moveable polymer ends are the rule
rather than the exception [4].

Following the standard recipes of classical statistical mechanics one finds
the Helmholtz free energy

$$F = -k_B T \ln Q_{2M}$$

$$\cong -2k_B T \{M \ln(CA/A_0) - M + M \ln M - \lambda_1 NM\} , \qquad (4.28)$$

where Stirling's formula was used. Now the force $p$ of steric repulsion per
unit area is given by

$$p = -\left(\frac{1}{A} \frac{\partial F}{\partial L}\right)_{N,M,T} \cong -2k_B TN \frac{M}{A} \frac{\partial \lambda_1}{\partial L} . \qquad (4.29)$$

Substitution of (4.24) into the last equation finally gives for the pressure
due to steric repulsion

$$p \cong k_B T\gamma \left(\frac{2NM}{AL}\right)^2 . \qquad (4.30)$$

This force increases proportional to $L^{-2}$ when the two cell membranes (or
colloidal particles) are brought together.

## References

[1]   P.G. de Gennes, *Scaling Concepts in Polymer Physics* (Cornell University Press, Ithaca, 1979).

[2]   R.J. Rubin, *J. Chem. Phys.* **43** (1965) 2392, **44** (1966) 2130, **51** (1969) 4681, **55** (1971) 4318; *J. Res. N.B.S.* **69B** (1965) 301, **70B** (1966) 237; *J. Math. Phys.* **8** (1967) 576.

[3]   F.W. Wiegel, *J. Physics* **A10** (1977) 299.

[4]   A.S. Perelson, C. DeLisi and F.W. Wiegel, eds., *Cell Surface Dynamics: Concepts and Models* (Marcel Dekker, New York, 1984).

[5]   P.G. de Gennes, *Rep. Prog. Phys.* **32** (1969) 187.

[6]   F.W. Wiegel, *Fluid Flow through Porous Macromolecular Systems*, Lecture Notes in Physics **121** (Springer, Berlin, 1980).

[7]   D.S. McKenzie, *Phys. Rep.* **27** (1976) 35.

[8]   C. Domb, *Adv. Chem. Phys.* **15** (1969) 229.

[9]   S.F. Edwards, *Proc. Phys. Soc.* **85** (1965) 613.

[10]  S.F. Edwards, in *Critical phenomena* (N.B.S., Washington, 1965) Misc. Publ. 273, p. 225.

[11]  K. Freed, *Adv. Chem. Phys.* **22** (1972) 1.

[12]  F.W. Wiegel, *Phys. Reports* **16** (1975) 57, subsection 3.3.

[13]  M. Muthukumar and B.G. Nickel, *J. Chem. Phys.* **80** (1984) 5839.

[14]  A.K. Dolan and S.F. Edwards, *Proc. R. Soc. Lond.* **A343** (1975) 427.

# IV. PATH INTEGRALS WITH TOPOLOGICAL CONSTRAINTS: POLYMER ENTANGLEMENTS, KNOTS AND LINKS

In several areas of theoretical physics, problems arise which, when formulated in the language of path integration, lead to integrals over collections of paths that are characterized by a certain global property. The simplest problems of this type arise when one considers the physical properties of entangled macromolecules, and it is the study of these problems which will be our concern in this chapter. In the next chapter we shall discuss the Aharonov-Bohm effect in quantum mechanics, where a similar situation occurs. We now turn to a discussion of polymer entanglement problems, following the method of Ref. 1.

## 4.1 The simple entanglement problem

In order to formulate the simplest problem in which a topologically constrained path integral plays a role, recall Eqs. (3.1, 12a, b) which state that the number ($Q$) of configurations of a macromolecule of length $Nl$, with endpoints at $\mathbf{r}_0$ and $\mathbf{r}_N$ is given by the Wiener integral

$$Q(\mathbf{r}_N, N \mid \mathbf{r}_0, 0) = \int_{\mathbf{r}_0, 0}^{\mathbf{r}_N, N} \exp\left\{-\frac{3}{2l^2} \int_0^N \left(\frac{d\mathbf{r}}{d\nu}\right)^2 d\nu\right\} d[\mathbf{r}(\nu)] \ . \quad (1.1)$$

Suppose one wants to answer the question: how many of these configurations wind themselves exactly $n$ times around some fixed curve $D$? To answer this question one has to proceed in two steps. In the first place one has to find some functional $L[\mathbf{r}(\nu)]$ of the polymer configuration which has the property

$$L[\mathbf{r}(\nu)] = n \ , \quad \text{if } \mathbf{r}(\nu) \text{ winds itself } n \text{ times around } D,$$

$$L[\mathbf{r}(\nu)] = 0 \ , \quad \text{otherwise.} \tag{1.2}$$

In terms of this functional the number $(Q_n)$ of $n$ times entangled configurations is given by

$$Q_n(\mathbf{r}_N, N \mid r_0, 0) = \int_{\mathbf{r}_0, 0}^{\mathbf{r}_N, N} \delta(n - L[\mathbf{r}(\nu)])$$

$$\times \exp\{-\frac{3}{2l^2} \int_0^N \left(\frac{d\mathbf{r}}{d\nu}\right)^2 d\nu\} \, d[\mathbf{r}(\nu)] \ . \tag{1.3}$$

A slightly more general situation arises if an external force, with a scalar potential $V(\mathbf{r})$, acts on the end point of each link in the polymer. In this case an extra Boltzmann factor

$$\exp\{-\beta \int_0^N V(\mathbf{r}(\nu)) \, d\nu\} \ , \tag{1.4a}$$

$$\beta = (k_B T)^{-1} \ , \tag{1.4b}$$

has to be included in the definition of the configuration sum, and hence the sum over the $n$-times entangled configurations is given by the path integral

$$Q_n(\mathbf{r}_N, N \mid r_0, 0) = \int_{\mathbf{r}_0, 0}^{\mathbf{r}_N, N} \delta(n - L) \exp\left\{ -\frac{3}{2l^2} \int_0^N \left(\frac{d\mathbf{r}}{d\nu}\right)^2 d\nu \right.$$

$$\left. - \beta \int_0^N V \, d\nu \right\} d[\mathbf{r}(\nu)] \ . \tag{1.5}$$

The second step in the calculation of course consists of the evaluation of this Wiener integral. This can be done in a variety of ways which have been reviewed in [2].

Before we can discuss the various methods for the evaluation of (1.5) we need an explicit formula for the functional $L[\mathbf{r}(\nu)]$. For the case of a polymer entangled with a closed curve $D$ this problem was essentially solved in the last century by the founding fathers of the theory of the electromagnetic field; their results were adapted to polymer physics by Fuller [3]. The idea essentially is to consider the vector field

$$\mathbf{F}(\mathbf{r}) = (4\pi)^{-1} \oint_D | \mathbf{r} - \mathbf{s} |^{-3} \, d\mathbf{s} \times (\mathbf{r} - \mathbf{s}) \, , \qquad (1.6)$$

where $d\mathbf{s}$ denotes the vectorial line element of $D$ and $\times$ the outer product of two vectors. It is shown in any textbook on electromagnetism [4] that this vector field has the following three properties

$$\operatorname{div} \mathbf{F} = 0 \, , \qquad (1.7)$$

$$\operatorname{curl} \mathbf{F} = 0 \quad \text{if } \mathbf{r} \notin D \, , \qquad (1.8a)$$

$$\operatorname{curl} \mathbf{F} = \infty \quad \text{if } \mathbf{r} \in D \, . \qquad (1.8b)$$

The curl of $\mathbf{F}$ diverges on $D$ in such a way that the line integral

$$\oint \mathbf{F} \cdot d\mathbf{r} = \oiint \operatorname{curl} \mathbf{F} \cdot d^2 \mathbf{S} = 1 \qquad (1.9)$$

for every small circle which winds arounds a line element of $D$ once, in such a direction that the vector which represents the line element and the direction of the circle are connected by the right-hand rule. From these properties it follows that

$$L[\mathbf{r}(\nu)] = \oint_C \mathbf{F} \cdot d\mathbf{r} = (4\pi)^{-1} \oint_C d\mathbf{r} \cdot \oint_D |\mathbf{r} - \mathbf{s}|^{-3} d\mathbf{s} \times (\mathbf{r} - \mathbf{s}) \, , \qquad (1.10)$$

where $C$ denotes the continuous curve consisting of the points $\mathbf{r}(\nu)$ with $0 \le \nu \le N$. This quantity is called the linking number, winding number or entanglement index of $C$ with respect to $D$.

The simplest method for the evaluation of (1.5) consists of substitution of the last equation, and use of the continuous representation for the delta function

$$\delta(x) = \int_{-\infty}^{\infty} \exp(2\pi k x i) \, dk \, . \qquad (1.11)$$

This leads to

$$Q_n(\mathbf{r}_N, N | \mathbf{r}_0, 0) = \int_{-\infty}^{\infty} \tilde{Q}_k \exp(2\pi k n i) \, dk \, . \qquad (1.12)$$

The Fourier transform is given by

$$\tilde{Q}_k = \int_{\mathbf{r}_0,0}^{\mathbf{r}_N,N} \exp\{- \int_0^N \left[ \frac{3}{2l^2} \left( \frac{d\mathbf{r}}{d\nu} \right)^2 + \beta V + 2\pi ki\, \mathbf{F} \cdot \frac{d\mathbf{r}}{d\nu} \right] d\nu\}\, d[\mathbf{r}(\nu)] \ .$$

(1.13)

This path integral has to be extended over all polymer configurations, irrespective of their winding number.

In order to proceed with the evaluation one completes the square in the exponential

$$\frac{3}{2l^2} \left( \frac{d\mathbf{r}}{d\nu} \right)^2 + 2\pi ki\, \mathbf{F} \cdot \frac{d\mathbf{r}}{d\nu} + \beta V = \frac{3}{2l^2} \left( \frac{d\mathbf{r}}{d\nu} + \frac{2}{3}\, \pi kl^2 i\mathbf{F} \right)^2$$

$$+ \frac{2}{3}\, \pi^2 l^2 k^2 \mathbf{F}^2 + \beta V \ . \qquad (1.14)$$

Now path integrals of the type

$$G_0(\mathbf{r}_N, N \mid \mathbf{r}_0, 0) = \int_{\mathbf{r}_0,0}^{\mathbf{r}_N,N} \exp\{-\frac{3}{2l^2} \int_0^N \left( \frac{d\mathbf{r}}{d\nu} + \frac{2}{3}\, \pi kl^2 i\mathbf{F} \right)^2$$

$$\times\, d\nu\}\, d[\mathbf{r}(\nu)]$$

(1.15)

formed the subject of Chapter II. Using property (1.7) and a generalization of (II.1.7) and (II.2.2) one finds that $G_0$ is the solution of

$$\left[ \frac{\partial}{\partial N} - \frac{l^2}{6}\triangle - \frac{2}{3}\, \pi kl^2 i\, \nabla \cdot \mathbf{F} \right] G_0(\mathbf{r}_N, N \mid \mathbf{r}_0, 0) = \delta(\mathbf{r}_N - \mathbf{r}_0)\delta(N) \ .$$

(1.16)

Following the theory of Section 1.4 it is straightforward to show that the additional terms $\frac{2}{3}\pi^2 l^2 k^2 \mathbf{F}^2 + \beta V$ in (1.14) lead to identical terms in the partial differential equation

$$\left[ \frac{\partial}{\partial N} - \frac{l^2}{6}\triangle - \frac{2}{3}\, \pi kl^2 i\, \nabla \cdot \mathbf{F} + \frac{2}{3}\, \pi^2 l^2 k^2 \mathbf{F}^2 + \beta V \right] \tilde{Q}_k$$

$$= \delta(\mathbf{r}_N - \mathbf{r}_0)\, \delta(N) \ . \qquad (1.17)$$

Hence $\tilde{Q}_k$, which is still a function of $\mathbf{r}_N$, $N$ and of $\mathbf{r}_0$, is the propagator of this equation and can therefore be expanded in its eigenfunctions

$$\tilde{Q}_k(\mathbf{r}_N, N \mid \mathbf{r}_0, 0) = \sum_a f_a(\mathbf{r}_N)\, f_a^*(\mathbf{r}_0)\, \exp(-\lambda_a N) \ , \qquad (1.18)$$

$$\left[ -\frac{l^2}{6}\Delta - \frac{2}{3}\pi k l^2 i \nabla\cdot\mathbf{F} + \frac{2}{3}\pi^2 l^2 k^2 \mathbf{F}^2 + \beta V \right] f_a(\mathbf{r}) = \lambda_a f_a(\mathbf{r}) \ . \quad (1.19)$$

For this expansion the eigenfunctions have to be made orthornormal. The boundary condition on the eigenfunctions have been discussed in Section 3.2; from (III. 2.7) we have

$$f_a(\mathbf{r}) = 0 \ , \qquad \mathbf{r} \in D \ . \quad (1.20)$$

It is also important to note that the operator $i \nabla\cdot\mathbf{F}$ is a Hermitian operator, which can be proved as follows. For any function $\phi$ which is smooth enough and vanishes at the hard walls of the system one has

$$i \nabla\cdot\mathbf{F} \, \phi = i\phi(\text{div } \mathbf{F}) + i \mathbf{F}\cdot\nabla\phi$$

$$= i \mathbf{F}\cdot\nabla\phi \quad (1.21)$$

because of property (1.7). Hence for any two such functions $\phi_1$ and $\phi_2$

$$(\phi_1, i \nabla\cdot\mathbf{F}\phi_2) = i \int \phi_1^* \, \nabla\cdot\mathbf{F}\phi_2 \, d^3r$$

$$= i \int \nabla\cdot\phi_1^* \mathbf{F}\phi_2 \, d^3r - i \int \phi_2 \mathbf{F}\cdot\nabla\phi_1^* \, d^3r$$

$$= (\phi_2, i \nabla\cdot\mathbf{F}\phi_1)^* \ , \quad (1.22)$$

where we used Gauss' theorem in the second line and (1.20) in the third. Because of the Hermitian character of all terms in the differential operator (1.19) all the eigenvalues $\lambda_a$ are real

The sum $Q_n(N)$ over the $n$-times entangled configurations of a closed polymer with an arbitrary location of its initial repeating unit is found by putting $\mathbf{r}_N = \mathbf{r}_0 = \mathbf{r}$ in (1.18) and integrating over all space. Using (1.12) we find for this quantity

$$Q_n(N) = \int_{-\infty}^{+\infty} \sum_a \exp\{-\lambda_a(k)N + 2\pi nki\} \, dk \ . \quad (1.23)$$

Of course, the full configuration sum $Q(N)$ over all the polymer configurations is found by integrating the last formula over all values $n$ of the winding number. As

$$\int_{-\infty}^{+\infty} \exp(2\pi nki) \, dn = 2\pi \, \delta(2\pi k) \quad (1.24)$$

this gives

$$Q(N) = \sum_a \exp\{-\lambda_a(0)N\} \ , \tag{1.25}$$

where the $\lambda_a(0)$ are to be solved from

$$\left[ -\frac{l^2}{6} \Delta + \beta V \right] f_a(\mathbf{r}) = \lambda_a(0) f_a(\mathbf{r}) \ . \tag{1.26}$$

What we have found in this way is a simple recipe for the calculation of the *a priori* probability $P_n(N)$ that a ring-shaped polymer will be $n$ times entangled with the curve $D$: this probability is given by the ratio

$$W_n(N) = \frac{Q_n(N)}{Q(N)} \ , \tag{1.27}$$

where the $Q_n(N)$ can be found, by way of Eq. (1.23), from the eigenvalues of Eq. (1.19). Depending on the form of the curve $D$—represented by the field $\mathbf{F}$—and on the explicit form of the interaction between the polymer and the curve—represented by $V$—there are now basically three possibilities:

(a)  The eigenvalue problem (1.19) has no bound states. In this case the summation over $a$ in (1.23) has to be extended over the whole spectrum.

(b)  The eigenvalue problem (1.19) has at least one bound state for all values of $k$. If $\lambda_0(k)$ denotes the ground state then for $N \gg 1$ the sum over $a$ will be dominated by the $a = 0$ term and one finds

$$Q_n(N) \cong \int_{-\infty}^{\infty} \exp\{-\lambda_0(k)N + 2\pi nki\} \, dk \ , \qquad (N \gg 1) \ . \tag{1.28}$$

(c)  The spectrum is continuous for some values of $k$, but a bound state occurs for other values of $k$.

In the remaining sections of this chapter we shall consider examples of cases (a) and (b). I do not know of an example of case (c) and am not quite sure if this case is a mathematical possibility at all.

## 4.2  Entanglement with a straight line

The entanglement problem which has received most attention in the literature is the special case in which $D$ is an infinitely long straight line, for example, the $z$-axis of a Cartesian frame of reference, and in which there is no interaction between the polymer and this line. The so-called simple entanglement problem is an example of case (a) of Section 4.1 and forms the subject of several papers [5–8 and III–6]. As its path integral solution has been reviewed in full detail by the author [1] we shall not again consider it here. Instead we shall consider an example of case (b) of Section 4.1, in which one has the same straight line geometry, but an interaction between a monomer at distance $r$ and the line with a potential $V(r)$ which will first be left general and later specialized to the case

$$V(r) = Cr^2 + \frac{D}{r^2} , \qquad (C>0, \quad D>0) . \qquad (2.1)$$

This interaction is repulsive at short distances and harmonically attractive at large distances, and therefore hopefully not too unrealistic.

The first step in the calculation consists of evaluating the line integral (1.6) for this geometry. The Cartesian $x$, $y$ and $z$ components of the vector field are

$$F_1(x,y,z) = \frac{-y}{2\pi(x^2+y^2)} , \qquad (2.2a)$$

$$F_2(x,y,z) = \frac{+x}{2\pi(x^2+y^2)} , \qquad (2.2b)$$

$$F_3(x,y,z) = 0 . \qquad (2.2c)$$

Using cylindrical coordinates $r$, $\theta$, $z$ around the line $D$ and substituting (2.2) into (1.19) one finds the eigenvalue problem

$$\left[ -\frac{l^2}{6} \left( \frac{\partial^2}{\partial z^2} + \frac{\partial^2}{\partial r^2} + \frac{1}{r}\frac{\partial}{\partial r} + \frac{1}{r^2}\frac{\partial^2}{\partial \theta^2} \right) - \frac{i}{3} kl^2 \frac{1}{r^2}\frac{\partial}{\partial \theta} \right.$$

$$\left. + \frac{l^2}{6}\frac{k^2}{r^2} + \beta V(r) \right] f_a = \lambda_a f_a . \qquad (2.3)$$

The dependence of the eigenfunctions on $z$ and $\theta$ has the form

$$f_a \equiv f_{p,q,m} = (2\pi L)^{-1/2} \exp(ipz + iq\theta) \, \phi_m(r) , \qquad (2.4)$$

where $q$ is an integer because of the periodic boundary conditions which have to be imposed on $\theta$ in the interval $(-\pi, +\pi)$. The quantity $p$ equals

$$p = \frac{2\pi}{L} \times \text{integer} \qquad (2.5)$$

if periodic boundary conditions are imposed on $z$ in the interval $(0, L)$; at the end of the calculation we shall take the limit $L \to \infty$. The equation for the $r$-dependent part $\phi_m$ of the eigenfunctions is

$$\left[ \frac{d^2}{dr^2} + \frac{1}{r}\frac{d}{dr} - \frac{(q+k)^2}{r^2} - \frac{6\beta}{l^2} V(r) + \frac{6\Lambda_m}{l^2} \right] \phi_m(r) = 0 \ , \qquad (2.6a)$$

$$\frac{6\Lambda_m}{l^2} \equiv \frac{6\lambda_a}{l^2} - p^2 \ , \qquad (2.6b)$$

$$\phi_m(0) = \phi_m(\infty) = 0 \ . \qquad (2.6c)$$

Note that $\Lambda_m$ and $\phi_m(r)$ will be functions of the sum $q + k$.

At this point it is of interest to substitute (2.4) into the bilinear expansion (1.18). This gives

$$\tilde{Q}_k(z_N, r_N, \theta_N, N \mid z_0, r_0, \theta_0, 0) = (2\pi L)^{-1} \sum_{p,q,m} \phi_m(r_N) \, \phi_m^*(r_0)$$

$$\times \exp\{ip(z_N - z_0) + iq(\theta_N - \theta_0) - \frac{1}{6} l^2 p^2 N - \Lambda_m N\} \ . \qquad (2.7)$$

It is straightforward to explicitly perform the summation over $p$; in the limit $L \to \infty$ one finds

$$(2\pi L)^{-1} \sum_p \exp\{ip(z_N - z_0) - \frac{1}{6} l^2 p^2 N\}$$

$$= \frac{1}{4\pi^2} \left( \frac{6\pi}{Nl^2} \right)^{1/2} \exp\left\{ -\frac{3(z_N - z_0)^2}{2Nl^2} \right\} \ . \qquad (2.8)$$

Using (1.12) one therefore has the following intermediate result

$$Q_n(z_N, r_N, \theta_N, N \mid z_0, r_0, \theta_0) = \frac{1}{4\pi^2} \left( \frac{6\pi}{Nl^2} \right)^{1/2} \exp\left\{ -\frac{3(z_N - z_0)^2}{2Nl^2} \right\}$$

$$\times \int_{-\infty}^{+\infty} dk \sum_{q,m} \phi_m(r_N) \, \phi_m^*(r_0) \exp\{iq(\theta_N - \theta_0) + 2\pi kni - \Lambda_m(q + k)N\} \ . \qquad (2.9)$$

It has been noted after (2.6) that the $\phi_m(r)$ as well as the $\Lambda_m$ will be functions of $q + k$. Hence, in order to calculate the configuration sum $Q_\mu(z, r, \theta, N \mid z, r, \theta, 0)$ of a ring-shaped macromolecule with both of its endpoints fixed at $(z, r, \theta)$ and $\mu$ times entangled with the straight line, we call $q + k = k'$, drop the prime on $k'$ again and find for the right-hand side of (2.9) the expression

$$Q_\mu(z, r, \theta, N \mid z, r, \theta, 0) = \frac{1}{4\pi^2} \left( \frac{6\pi}{Nl^2} \right)^{1/2} \int_{-\infty}^{+\infty} dk \sum_{q,m}$$

$$\times |\phi_m(r)|^2 \exp\{2\pi(k - q)\mu i - \Lambda_m(k)N\} \ , \qquad (2.10)$$

where $\Lambda_m$ and $\phi_m$ are now defined by (2.6) with $q + k$ replaced by $k$. The summation over all integer values of $q$ can be performed with the formula

$$\sum_{q=-\infty}^{+\infty} \exp(-2\pi q\mu i) = \sum_{q=-\infty}^{+\infty} \delta(\mu - n) \ , \tag{2.11}$$

which shows that $Q_\mu$ will only be different from zero if the entanglement index $\mu$ equals an integer $n$. This can be expressed in the somewhat symbolic formula

$$\int_{n-0}^{n+0} Q_\mu(z, r, \theta, N \mid z, r, \theta, 0) \, d\mu$$

$$= \frac{1}{4\pi^2} \left( \frac{6\pi}{Nl^2} \right)^{1/2} \int_{-\infty}^{+\infty} dk \sum_m | \phi_m(r) |^2 \exp\{2\pi kni - \Lambda_m(k)N\} \ . \tag{2.12}$$

Usually we shall denote the left-hand side of this equation simply by $Q_n(z, r, \theta, N \mid z, r, \theta, 0)$, but it should be kept in mind that this is a discrete notation for an entanglement index that can take continuous values.

Using the same discrete notation the configuration sum $Q_n(N)$ of a ring-shaped macromolecule which is $n$ times entangled with a straight line, but without any constraints otherwise, follows by integrating (2.12) over all $r$, $\theta$ space and $z$ from 0 to $L$. This gives

$$Q_n(N) = \frac{L}{4\pi^2} \left( \frac{6\pi}{Nl^2} \right)^{1/2} \int_{-\infty}^{+\infty} \sum_m \exp\{-\Lambda_m(k)N + 2\pi nki\} \, dk \ , \tag{2.13}$$

where $\Lambda_m(k)$ is defined by (2.6) with $q + k$ replaced by $k$. This formula was first derived by the author in Ref. 8, using a different method. It is similar to (1.23) but with two of the three summations over the eigenvalues performed explicitly already.

Consider now the special case of an interaction of the form (2.1). The eigenvalues $\Lambda_m(k)$ have to be solved from

$$\left[ \frac{d^2}{dr^2} + \frac{1}{r} \frac{d}{dr} - \frac{k^2}{r^2} - \frac{6\beta}{l^2} \left( Cr^2 + \frac{D}{r^2} \right) + \frac{6\Lambda_m}{l^2} \right] \phi_m(r) = 0 \ . \tag{2.14}$$

As $V(r) \to \infty$ for $r \to \infty$ the spectrum of the $\Lambda_m$ will be discrete for any $k$. For the ground state one can try the ansatz

$$\phi_0(r) = \text{constant} \cdot r^\alpha \exp(-\gamma r^2) \ , \tag{2.15}$$

where $\alpha > 0$ and $\gamma > 0$ are adjustable real numbers; the value of the constant follows by normalization. Upon substitution of the ansatz into (2.14) one finds immediately that (2.15) is a solution provided

$$\alpha = \sqrt{k^2 + \frac{6\beta D}{l^2}} , \tag{2.16}$$

$$\gamma = \sqrt{\frac{3\beta C}{2l^2}} , \tag{2.17}$$

$$\Lambda_0 = \sqrt{\frac{2}{3} \beta C l^2} \left( 1 + \sqrt{k^2 + \frac{6\beta D}{l^2}} \right) . \tag{2.18}$$

Hence (2.15) is an eigenfunction, and, as this eigenfunction has no node for $0 < r < \infty$, it has to be the ground state. For $N \gg 1$ Eq. (2.13) gives, for the sum over the $n$ times entangled configurations

$$Q_n(N) = \frac{L}{4\pi^2} \left( \frac{6\pi}{Nl^2} \right)^{1/2} \exp\left( -Nl\sqrt{\frac{2}{3} \beta C} \right) \int_{-\infty}^{+\infty} \exp\{\psi(k)\} \, dk , \tag{2.19a}$$

$$\psi(k) \equiv 2\pi k n i - Nl\sqrt{\frac{2}{3} \beta C} \left( k^2 + \frac{6\beta D}{l^2} \right) . \tag{2.19b}$$

In the general case, in which $C$ and $D$ are both non-zero, the integral (2.19) cannot be evaluated analytically. For $N \gg 1$ one can find an asymptotic expression with the saddle-point method. For $|n| \ll N$ this gives

$$Q_n(N) \cong \text{constant} \cdot \exp\left( -\frac{6\pi^2 n^2}{Nl^2} \sqrt{\frac{D}{C}} \right) , \qquad (|n| \ll N, N \gg 1) , \tag{2.20}$$

which after normalization leads to the entanglement probabilities

$$W_n(N) \cong \left( \frac{6\pi}{Nl^2} \right)^{1/2} \left( \frac{D}{C} \right)^{1/4} \exp\left( -\frac{6\pi^2 n^2}{Nl^2} \sqrt{\frac{D}{C}} \right) . \tag{2.21}$$

Note that the temperature does not appear in this result anymore.

The case $D = 0$, $V(r) = Cr^2$ corresponds to a harmonic interaction between the monomers and the line. In this case the integral (2.19) can be evaluated analytically with the result

$$Q_n(N) = \frac{L}{4\pi^2} \left( \frac{6\pi}{Nl^2} \right)^{1/2} \frac{2Nl\sqrt{\frac{2}{3} \beta C}}{\frac{2}{3} \beta C l^2 N^2 + 4\pi^2 n^2} \exp\left( -Nl\sqrt{\frac{2}{3} \beta C} \right) . \tag{2.22}$$

Using (1.27) and the summation formula

$$\sum_{n=-\infty}^{+\infty} \frac{1}{S^2 + 4\pi^2 n^2} = \frac{1}{2S} \coth\frac{1}{2} S \ , \tag{2.23}$$

to determine the normalization one finds, for the entanglement probabilities in thermal equilibrium,

$$W_n(N) = \left(\coth\frac{1}{2} Nl \sqrt{\frac{2}{3} \beta C}\right)^{-1} \frac{2Nl\sqrt{\frac{2}{3} \beta C}}{\frac{2}{3} \beta Cl^2 N^2 + 4\pi^2 n^2} \ . \tag{2.24}$$

This formula implies $<n> = 0$ but $<n^2> = +\infty$, which shows that in the absence of a hard core in the monomer-line interaction (modelled by the term $D/r^2$) very large values of the entanglement index are likely to occur.

### 4.3 The Green's function for the half-plane barrier

As we remarked in the beginning of Section 4.2, the entanglement problem for a polymer with a straight line with which it has no interaction has been solved with a variety of methods [1]. We therefore quote the result for the configuration sum over the $n$ times entangled configurations

$$Q_n(z_N, r_N, \theta_N, N \mid z_0, r_0, \theta_0) = \left(\frac{2}{3} \pi Nl^2\right)^{-3/2}$$

$$\times \exp\left\{-\frac{3(z_N - z_0)^2}{2Nl^2} - \frac{3(r_N^2 + r_0^2)}{2Nl^2}\right\}$$

$$\times \int_{-\infty}^{+\infty} I_{|k|}\left(\frac{3r_0 r_N}{Nl^2}\right) \exp[(\theta_N - \theta_0 + 2\pi n)ki] \, dk \ , \tag{3.1}$$

where the $I_\nu$ denote the modified Bessel functions

If one only looks at the projection of the chain on the $x$, $y$ plane one has a two-dimensional entanglement problem between a random walk with an average squared step length $\frac{2}{3}l^2$ and the origin of the plane. The configuration sum $q_n$ is found by integrating (3.1) over $z_N$. Replacing $l^2$ by $\frac{3}{2}l^2$ in order to give the projected random walk an average squared step length equal to $l^2$, one finds for the configuration sum over the $n$ times entangled configurations

$$q_n(r_N,\ \theta_N,\ N\mid r_0,\ \theta_0) = (\pi N l^2)^{-1} \exp\left\{-\frac{(r_N^2 + r_0^2)}{Nl^2}\right\}$$

$$\times \int_{-\infty}^{+\infty} I_{|k|}\left(\frac{2r_0 r_N}{Nl^2}\right) \exp[(\theta_N - \theta_0 + 2\pi n)ki]\ dk\ .\quad (3.2)$$

This result will now be used to derive the Green's function for a Brownian particle which diffuses through space in the presence of a half-plane barrier (or in a plane in the presence of a half-line). For the sake of convenience we write all equations down for the two-dimensional problem; their generalization to three dimensions is quite straightforward.

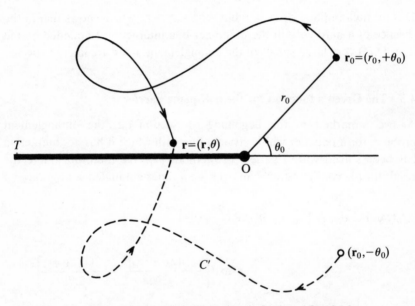

Fig. 4.1.   Basic geometry for the discussions in Sections 4.2 and 4.3.

The basic geometry is drawn in Fig. 4.1. We introduce the auxiliary variable $\tau_0$ with the dimensions of time and take the limit $N\to\infty, \tau_0\to 0$, $l\to 0$ such that $N\tau_0 = t$ and $l^2/4\tau_0 = D$ remain fixed. In this limit the polymer configurations turn into the paths of a particle which performs Brownian motion with a diffusion coefficient $D$ and the differential equation for the functions $q_n$ of Eq. (3.2)

$$\frac{\partial q_n}{\partial N} = \frac{l^2}{4}\,\triangle q_n\ ,\qquad\qquad (3.3)$$

turns into the diffusion equation

$$\frac{\partial q_n}{\partial t} = D \triangle q_n \ ,$$ (3.4)

with the initial condition

$$\lim_{t \to 0} q_n = \delta(\mathbf{r} - \mathbf{r}_0) \quad \text{if } n = 0,$$ (3.5a)

$$= 0 \qquad \text{if } n = \pm 1, \pm 2, \pm 3, \ldots$$ (3.5b)

Hence the *a priori* probability density for a Brownian motion path to be $n$ times entangled with the origin is found from (3.2)

$$q_n(r, \theta, t) = (4\pi Dt)^{-1} \exp\left\{-\frac{(r^2 + r_0^2)}{4Dt}\right\}$$

$$\times \int_{-\infty}^{+\infty} I_{|k|}\left(\frac{r_0 r}{2Dt}\right) \exp[(\theta - \theta_0 + 2\pi n)ki] \, dk \ ,$$ (3.6)

where $-\pi < \theta, \theta_0 < +\pi$ and $n = 0, \pm 1, \pm 2, \ldots$

An independent check of this formula consists of verifying that the sum of the probabilities adds up to a properly normalized Gauss distribution. Summing (3.6) over all integer values of $n$ and using the summation formula

$$\sum_{n=-\infty}^{+\infty} \exp(2\pi nki) = \sum_{m=-\infty}^{+\infty} \delta(k - m)$$ (3.7)

one finds for the total probability density

$$c(r, \theta, t) = \sum_{n=-\infty}^{+\infty} q_n(r, \theta, t)$$ (3.8)

$$= (4\pi Dt)^{-1} \exp\left\{-\frac{(r^2 + r_0^2)}{4Dt}\right\} \sum_{m=-\infty}^{+\infty} I_{|m|}\left(\frac{r_0 r}{2Dt}\right) \exp[im(\theta - \theta_0)] \ .$$

Using the generating function for the modified Bessel functions (Eq. (9.6.34) of Ref. 9) one then finds

$$c(r, \theta, t) = (4\pi Dt)^{-1} \exp\left[-\frac{\{r^2 + r_0^2 - 2rr_0 \cos(\theta - \theta_0)\}}{4Dt}\right]$$ (3.9)

which is indeed the properly normalized Gauss distribution for diffusion in the plane.

The formula (3.6) can be used to derive various interesting results, as has been shown recently in Ref. 10 and 11 which we follow in this section. First of all, one can study the properties of Brownian-motion paths in the plane which differ from free Brownian paths through an extra weight factor $(-1)$ for every crossing of the branch line $T$; cf. Fig. 4.1 and Ref. 12. The total weight of these paths is

$$p(r, \theta, t) = \sum_{n=-\infty}^{+\infty} (-1)^n q_n(r, \theta, t) \ . \tag{3.10}$$

Substituting (3.6) and using (3.7) one finds

$$p(r, \theta, t) = (2\pi Dt)^{-1} \exp\left\{-\frac{(r^2 + r_0^2)}{4Dt}\right\}$$

$$\times \sum_{m=0}^{\infty} I_{m+1/2}\left(\frac{r_0 r}{2Dt}\right) \cos[(m + \tfrac{1}{2})\,(\theta - \theta_0)] \ . \tag{3.11}$$

The identity

$$\sum_{m=0}^{\infty} I_{m+1/2}(z) \cos(m + \tfrac{1}{2})\,\phi = \frac{1}{2} \exp(z \cos \phi)\, \mathrm{erf}\left(\sqrt{2z}\, \cos\frac{\phi}{2}\right) \tag{3.12}$$

may be used to express the series in Eq. (3.11) in terms of the error function

$$\mathrm{erf}(z) = \frac{2}{\sqrt{\pi}} \int_0^z e^{-t^2}\, dt \ . \tag{3.13}$$

A proof of (3.12) can be found in [10]. As a result one finds, for the total weight of these Brownian-motion paths with alternating signs, a simple closed-form expression

$$p(r, \theta, t) = (4\pi Dt)^{-1} \exp\left(-\frac{R^2}{4Dt}\right) \mathrm{erf}\left(\sqrt{\frac{r_0 r}{Dt}}\, \cos\frac{1}{2}\,(\theta - \theta_0)\right) , \tag{3.14}$$

where

$$R^2 = r^2 + r_0^2 - 2rr_0\cos(\theta - \theta_0) \tag{3.15}$$

is the square of the distance between the initial point $(r_0, \theta_0)$ and the final point $(r, \theta)$ of the walks.

The total weight (3.14) is a sum over free Brownian-motion paths in which those crossing the branch line an odd number of times have an extra weight factor $(-1)$, and those which cross the branch line an even number of times have an extra weight factor $(+1)$. Let us call these paths "odd" paths and "even" paths, respectively, and let their total contributions, without the extra weight factors, be denoted by $O(r, \theta, t \mid r_0, \theta_0)$ and $E(r, \theta, t \mid r_0, \theta_0)$. Since

$$E(r, \theta, t \mid r_0, \theta_0) - O(r, \theta, t \mid r_0, \theta_0) = p(r, \theta, t) \ , \tag{3.16}$$

$$E(r, \theta, t \mid r_0, \theta_0) + O(r, \theta, t \mid r_0, \theta_0) = (4\pi Dt)^{-1} \exp\left(-\frac{R^2}{4Dt}\right) \ , \tag{3.17}$$

the separate contributions $E$ and $O$ of the even and odd Brownian motion paths can be solved after substitution of (3.14). One finds

$$E(r, \theta, t \mid r_0, \theta_0) = (8\pi Dt)^{-1} \exp\left(-\frac{R^2}{4Dt}\right)$$

$$\times \left\{1 + \mathrm{erf}\left(\sqrt{\frac{r_0 r}{Dt}} \cos\frac{1}{2}(\theta - \theta_0)\right)\right\} \ , \tag{3.18}$$

$$O(r, \theta, t \mid r_0, \theta_0) = (8\pi Dt)^{-1} \exp\left(-\frac{R^2}{4Dt}\right)$$

$$\times \left\{1 - \mathrm{erf}\left(\sqrt{\frac{r_0 r}{Dt}} \cos\frac{1}{2}(\theta - \theta_0)\right)\right\} \ . \tag{3.19}$$

Another interesting result which can be derived as a corollary of (3.6) is the analytical solution of Carslaw's heat conduction problem. In 1899 Carslaw [13, 14] solved the following two-dimensional heat conduction problem for the temperature $U(x, y, t)$:

$$\frac{\partial U}{\partial t} = D\triangle U \ , \qquad (-\infty < x, y < +\infty, t > 0) \ , \tag{3.20a}$$

$$U(x, y, 0) = \delta(x - x_0)\,\delta(y - y_0) \ , \tag{3.20b}$$

$$U(x, 0, t) = 0 \qquad (x < 0, t > 0) \ . \tag{3.20c}$$

The Dirichlet boundary condition (3.20c) can also be replaced by the Neumann boundary condition

$$\frac{\partial U}{\partial y}(x, 0, t) = 0 \ , \qquad (x < 0, \qquad t > 0) \ . \tag{3.20d}$$

Using our previous results we can immediately write down the solution of

Carslaw's problem. Indeed, in polar coordinates the solution is

$$U(r, \theta, t) = E(r, \theta, t \mid r_0, \theta_0) \mp O(r, \theta, t \mid r_0, -\theta_0) \; , \qquad (3.21)$$

where the $-$ sign applies to the Dirichlet boundary condition and the $+$ sign to the Neumann boundary condition. The correctness of the last formula is due to the following three observations:

(a)   The functions $E$ and $O$ are sums of functions $q_n$ which satisfy (3.4); hence the right-hand side of (3.21) is a solution of (3.20a).

(b)   In the limit $t \to 0$, $E$ tends to $\delta(\mathbf{r} - \mathbf{r}_0)$ and $O$ vanishes: hence the initial condition (3.20b) is satisfied.

(c)   In the limit $\theta \to \pm\pi$ there exists a one-to-one correspondence between the even (odd) paths from $(r_0, \theta_0)$ to $(r, \theta)$ and the odd (even) paths from $(r_0, -\theta_0)$ to $(r, \theta)$, because these paths can be mapped into each other by reflection with respect to the $x$-axis (see Fig. 4.1). Hence $E - O$ will vanish for $\theta = \pm\pi$ and thereby fulfill the boundary condition (3.20c); for the same reason $E + O$ fulfills (3.20d).

Combination of the last formula with (3.18, 19) gives Carslaw's original formula

$$U(r, \theta, t) = (8\pi Dt)^{-1} \exp\left(-\frac{R^2}{4Dt}\right) \left\{ 1 + \operatorname{erf}\left(\sqrt{\frac{r_0 r}{Dt}} \cos\frac{1}{2}(\theta - \theta_0)\right) \right\}$$

$$(3.22)$$

$$\mp (8\pi Dt)^{-1} \exp\left(-\frac{R_1^2}{4Dt}\right) \left\{ 1 - \operatorname{erf}\left(\sqrt{\frac{r_0 r}{Dt}} \cos\frac{1}{2}(\theta + \theta_0)\right) \right\} \; ,$$

where

$$R_1^2 = r^2 + r_0^2 - 2r\,r_0 \cos(\theta + \theta_0) \qquad (3.23)$$

is the square of the distance between $(r_0, -\theta_0)$ and $(r, \theta)$.

Finally we observe that a trivial modification of (3.22) leads to the Green's function of the Schrödinger equation

$$i\hbar\,\frac{\partial\psi}{\partial t} = -\frac{\hbar^2}{2m}\,\triangle\psi \qquad (3.24)$$

for a free particle of mass $m$ moving in a plane with a half-line barrier. Indeed, through the substitution $D \to \hbar^2/2m$ and $t \to it/\hbar$ one immediately finds

$$\psi(r, \theta, t) = (2\,\pi i\hbar t/m)^{-1} \exp\left(\frac{imR^2}{2\hbar t}\right) F\left(\sqrt{\frac{2mr_0 r}{\hbar t}} \cos\frac{1}{2}(\theta - \theta_0)\right) \qquad (3.25)$$

$$\mp (2\pi i\hbar t/m)^{-1} \exp\left(\frac{imR_1^2}{2\hbar t}\right) F\left(-\sqrt{\frac{2mr_0 r}{\hbar t}} \cos\frac{1}{2}(\theta + \theta_0)\right)$$

in terms of the Fresnel integral

$$F(z) = \frac{1}{\sqrt{i\pi}} \int_{-\infty}^{z} e^{it^2} \, dt \; . \tag{3.26}$$

Here the upper (lower) sign is for Dirichlet (Neumann) boundary conditions on the negative $x$-axis. The same formula was recently derived by Schulman [15, 16]. His derivation is based on the semi-classical approximation to the path integral, using a single intermediate time, which rather surprisingly yields the exact solution. Our derivation of (3.22, 25) as the difference of constrained path integrals over even and odd paths, is another indication that the solution of Carslaw's problem, when expressed as a sum over Brownian motion paths, has a peculiar simplicity.

The Green's function (3.25) describes the diffraction of the de Broglie waves at the edge of the half plane. Further applications of path integrals to diffraction have been discussed by Lee [17], Dashen [I–39] and especially in the monograph by Schulman [I–35].

## 4.4 The statistical physics of knots

The first three sections of this chapter were devoted to the simplest problems in which the topology of polymer configurations plays a role: the first polymer is represented by some fixed curve $D$, the second polymer winds itself $n$ times around $D$. In many cases the exact solution could be found. We now turn to much more complicated problems which have to do with the entanglement of a single, closed polymer with itself. Before proceeding to a brief discussion of the physical relevance of these questions we must clarify the model and some of the concepts involved.

The polymer will be described with the free random walk model of Section 3.1; a closed polymer is thus represented by a three-dimensional polygon consisting of $N$ segments of length $l$. Two configurations $C_1$ and $C_2$, of a closed polymer will be called topologically equivalent if $C_1$ can be continuously deformed into $C_2$ in such a way that intersection of segments never occurs in any of the intermediate configurations. It is clear that the set of all configurations can now be divided in an infinite number of classes of topologically equivalent configurations (knots). A closed polymer will be called non-self-entangled if its configurations are equivalent to a circle; this is the trivial knot. If a macromolecular system consists of two or more closed polymers its configurations can again be divided into a discrete set of classes (links), each of which consists of topologically equivalent configurations. In such cases one makes the transition between one class and another by "forcing" an intersection between two segments on the same or different polymers.

Knots and links have appealed to the human imagination since the beginning of history. For example, the world picture of the ancient Mayas resembled a woven structure consisting of complicated knots and links [18]. The Incas used knotted strings (quipus) as mnemonic devices [19]. Links of three rings appear as mandalas in Tibetan Buddhism, again somehow representing the totality of being. This list may be extended considerably.

At the end of the 19th century the idea of knots entered into theoretical physics when Kelvin tried to interprete atoms as knots tied in the vortex lines which were then believed to lace the ether. Although these ideas were superseded by later developments in atomic physics the interest of mathematicians was aroused. They developed the mathematical discipline of knot theory, an exposition of which can be found, for example, in the monographs of Fox [20], Reidemeister [21], Crowell and Fox [22] and Rolfsen [23].

Knot theory came back to the experimental sciences around 1962, when Delbrück asked which fraction of the configurations of a closed polymer will have no knot [24]. Here, the total number of closed polymer configurations (with one monomer fixed in space!) will grow like

$$Q(N) \cong A\lambda^N N^{-a} , \qquad (N \gg 1) , \qquad (4.1)$$

where $A$ and $\lambda$ are positive constants. The exponent $a$ is universal (i.e., independent of the details of the polymer) and has the value $a = 3/2$ for a free random walk model and $a \cong 7/4$ for a self avoiding random walk model [25]. Delbrück conjectured that the number of unknotted closed polymer configurations grows, for large $N$, like

$$Q_0(N) \cong A_0 \lambda_0^N N^{-a_0} \qquad (4.2)$$

with different constants $A_0$, $\lambda_0$, $a_0$. Taking the ratio of the last two equations we find that the *a priori* probability $\zeta_N$ that a closed polymer ring will be unknotted (i.e., that it will be topologically equivalent to a circle) is given by

$$\zeta_N \cong C\mu^N N^\alpha , \qquad (N \gg 1) , \qquad (4.3)$$

where $C = A_0/A$, $\mu = \lambda_0/\lambda$ and $\alpha = a - a_0$.

The physical significance of this result is quite interesting. For example, the author has shown in [1] that the exponent $\alpha$ is related to the order (sharpness) of the helix-coil transition in double-stranded DNA molecules. Another exampled is the viscosity of a concentrated solution of linear (non-closed) polymers. When such a solution is subjected to shear flow those polymers, which are mutually entangled, have to be pulled apart; this extra force leads to an additive term in the viscosity which is of a "topological" nature. The physics of these phenomena has been studied especially by de Gennes, whose monograph should be required reading for every serious student of the theory of polymer systems [26].

In this and the next sections we shall study some of the statistical questions related to knots from the point of view of a theoretical physicist. If one wants to form an idea about the various knots which are possible, the first step usually consists of looking at the steadily improving tables of knots (and links) which have been published by the mathematicians [27–31, 21, 23]. In these tables the knots are represented by their projections onto a plane and ordered by an increasing number of double points in these projections. Even the most extensive table available today, due to Rolfsen [23], contains only a few hundred knots with less than a dozen double points. Hence, if one wants, for example, to determine the *a priori* probability that a very long closed polymer ring will be unknotted, these tables of knots and rings are of no use. This is especially true of the very long ($N \sim 10^3$ to $10^4$) random walks we are interested in because, from a physical point of view, they are good models for real macromolecules and, from a mathematical point of view, they are good approximations of the paths in a path integral.

Next, having found that the tables of knots are of no use for our purpose (they are marvellous works of art otherwise!) one might try to develop the statistical physics of knots analytically. In order to do this one needs an analytical invariant for a knot; i.e., if $r(\lambda)$, $0 \le \lambda \le L$, is some closed differentiable curve in space, of total length $L$, one needs some analytical expression in terms of $r(\lambda)$ the value of which is: (1) constant in every topological class; (2) different for different topological classes. However, up till now no analytical invariant for a knot has been found (cf. the brief discussion in Section V.D of Ref. 1). This makes the analytical solution of problems related to knots an almost hopeless task. Note the difference between these self entanglements (knots) for which no analytical invariant is known, and mutual-entanglements (links) for which the linking number (1.10) plays the role of an approximate analytical invariant. Hence, problems related to links can sometimes be solved analytically (as we did in Sections 4.1–4.3) but for problems related to the statistics of knots an analytical solution is in the crystal ball.

In view of these difficulties in developing the statistical physics of knots analytically there is a need for reliable numerical studies. As numerical enumerations have proved their use in other parts of physical theory one should expect them to be useful in this context too. These enumerations should have the following desirable properties: (1) closed polymer configurations have to be generated with the correct *a priori* probabilities; (2) given a configuration its topological properties should be determined in an appropriate way; (3) in order to achieve adequate statistical accuracy the number of polymer configurations should be large, which imposes the necessity for efficient algorithms. In the next two sections we shall discuss various aspects of the numerical approach, following Ref. 32 and 33. This approach to knot theory can properly be called numerical hammagraphy (from the Greek τό 'άμμα, meaning *the knot*).

### 4.5  Applicability and properties of Alexander polynomials

The characteristic feature of a knot is its specific way of self-entanglement, hence knots are classified according to this feature. The orthogonal projection of a knot onto a plane forms the starting point for its classification with the Alexander polynomial. In general this projection will show double points, some of which can be removed by a continuous deformation of the curve (without self-intersections, as described above!). The first, and crudest, characterization of the knot is simply the minimum number of these double points. In the existing tables of knots the knots are ordered by an increasing number of irreducible double points and illustrated by simple pictures of their plane projections.

We now introduce the concept of a "loop". A loop is a non-closed knot, that is an open, continuous curve in three-dimensional space. Thus, a loop can be obtained from a knot by omitting a section of the knot *which does not contain any double points*. Conversely, given a loop, one can form a

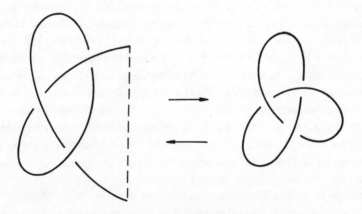

Fig. 4.2.   Example of a loop related to the trefoil knot.

knot by connecting the end points of the loop in such a way that no new double points are created (cf. Fig. 4.2).

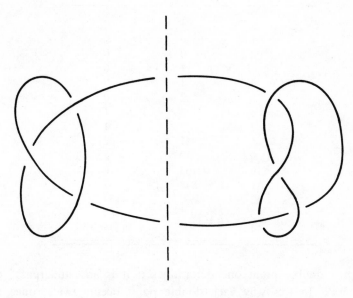

Fig. 4.3.   A non-prime knot which consists of two prime loops.

If a knot can be interpreted as a series of loops, as shown in Fig. 4.3, then this knot will be called non-prime. A prime knot cannot be decomposed in this way. Because of the one-to-one correspondence between knots and loops the loops too can be divided into prime loops and non-prime loops. Obviously, a table of knots need only contain the prime knots. In Table 4.1 we give the number $M(n)$ of prime knots and the number $F(n)$ of non-prime knots as a function of the minimum number $n$ of double points.

Although the projection of a knot onto a plane is useful as a visual representation, it is less appropriate as a starting point for quantitative considerations. For this purpose we need an algebraic invariant, i.e., some algebraic expression which is invariant for deformations of the type quoted above. It would also be useful if this algebraic expression characterized the knot in a 1-to-1 way. For this purpose the Alexander polynomial is generally accepted; some caveats will be discussed shortly.

The construction of the Alexander polynomial for a given knot starts with an arbitrary plane projection with a minimal number of double points. Starting from an arbitrary point (which should not be a double point) we trace the curve in an arbitrary, but fixed direction. Each time one passes

Table 4.1.  The number of prime (*M*) and non-prime (*F*) knots with *n* double points.

| $n$ | $M(n)$ | $F(n)$ |
|---|---|---|
| 0 | 1 | 0 |
| 1 | 0 | 0 |
| 2 | 0 | 0 |
| 3 | 1 | 0 |
| 4 | 1 | 0 |
| 5 | 2 | 0 |
| 6 | 3 | 1 |
| 7 | 7 | 1 |
| 8 | 21 | 2 |
| 9 | 49 | 5 |
| 10 | 166 | 10 |
| 11 | 548 | 37 |
| 12 | – | 154 |
| 13 | – | 484 |
| 14 | – | 1115 |

through a double point one determines if it is an "underpass" or an "overpass". In this way each double point occurs twice, once as an underpass and once as an overpass. We number the various underpasses in the order in which they occur. A part of the curve between two successive underpasses will be called a bridge. To a bridge one assigns the number of the underpass at which it ends.

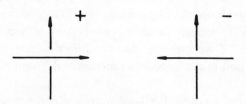

Fig. 4.4.  The two possibilities for crossings.

For a knot the projection of which has *n* double points we define an $n \times n$ matrix (the knot matrix) as follows: row $k$ corresponds to underpass $k$. Now one can classify underpass $k$ as belonging to type (+) if the corresponding bridge crosses from the left, or to type (−) if the corresponding bridge crosses from the right (Fig. 4.4). Suppose underpass $k$ is of type (+). Let $i$ denote the number of the corresponding bridge. If $i = k$ or $i = k + 1$ the elements of the knot matrix in row. $k$ are $a_{k,k} = -1$, $a_{k,k+1} = +1$ and the others vanish. If $i \neq k$ and $i \neq k + 1$ the elements of row

$k$ are $a_{k,k}=+1$, $a_{k,k+1}=-t$, $a_{k,i}=t-1$ and the others vanish. Next consider the case where underpass $k$ is of the type $(-)$, crossing with bridge $i$. If $i=k$ or $i=k+1$ we define $a_{k,k}=-1$, $a_{k,k+1}=+1$ and the other elements of row $k$ vanish. If $i \neq k$ and $i \neq k+1$ the elements of row $k$ of the knot matrix are $a_{k,k}=-t$, $a_{k,k+1}=+1$, $a_{k,i}=t-1$ and the others vanish (it should be mentioned that $k+1$ means 1 for $k=n$). This completes the recipe for the construction of the knot matrix.

In order to obtain the Alexander polynomial we choose an arbitrary (!) $n-1$ by $n-1$ minor of the knot matrix and calculate its determinant. This determinant will be a polynomial in $t$. Multiply this polynomial with $\pm t^{-m}$, where $m$ is chosen in such a way that the lowest power of $t$ is 0, and where the $\pm$ sign is chosen such that the $t^0$ term will be positive. The resulting polynomial $\Delta(t)$ is called the Alexander polynomial. The reader might want to apply this recipe to the trefoil knot of Fig. 4.2 and to verify that $\Delta(t)=t^2-t+1$.

The following properties of the Alexander polynomial should be noted; for their formal proof the interested reader should consult Ref. 23.

(a)   When the terms of $\Delta(t)$ are ordered by decreasing powers of $t$ the coefficients are integers, alternate in sign and are palindromic (i.e., they are invariant under a reversal of order). As their number is always odd any Alexander polynomial can be characterized uniquely by $\frac{1}{2}(m+2)$ integers, where $m$ is the order of the polynomial. Therefore, in the existing tables of knots, $\Delta(t)$ is often represented by this sequence of integers. The polynomial of the trivial knot turns out to equal unity.

(b)   The Alexander polynomial $\Delta(t)$ is indeed a topological invariant.

(c)   If a knot is non-prime, for example a series of two prime loops, then its Alexander polynomial is the product of the two Alexander polynomials corresponding to the two constitutent loops (where, of course, these loops have to be closed separately into two knots in order to allow the determination of their Alexander polynomials). An interesting corollary of this property is the fact that two non-trivial knots with $\Delta(t) \neq 1$ can never be put in series such that they reduce to a trivial knot.

### 4.6   Limitations on the use of the Alexander polynomial

The Alexander polynomial is not sacred and we should note the following limitations of its use.

First, it should be kept in mind that by way of exception two topologically different knots can have the same Alexander polynominal! Important examples are those prime knots (different from the trivial knot) which have $\Delta(t) = 1$. Three such knots have been discussed in the literature; they were

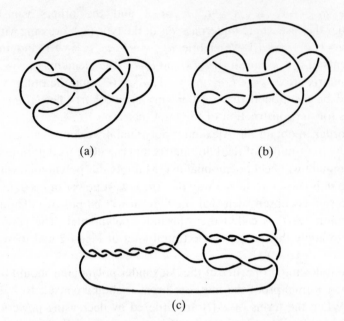

(a)                                        (b)

(c)

Fig. 4.5.   The knots of Kinoshita and Terasaka (a), Conway (b) and Seifert (c) have an
Alexander polynomial $\triangle(t) \equiv 1$.

discovered by Kinoshita and Terasaka [34, 35], Conway [31] and Seifert
[36]. They are drawn in Fig. 4.5a, b, c. The knots of Kinoshita and
Terasaka and of Conway are the simplest non-trivial knots with $\triangle(t) \equiv 1$;
their projections have 11 double points. As there are 795 prime knots with
$\leq 11$ double points the probability of finding a non-trivial knot with
$\triangle(t) \equiv 1$ is of the order $10^{-3}$. One should take this as an indication that
duplication of Alexander polynomials will be exceptional for more and
more complicated knots.

Second, we have noticed in Section 4.5 that the Alexander polynomial
of a knot which is the "product" of its constituent prime knots is the
product of the Alexander polynomials of those prime knots. Unfortunately, the inverse need not be true: sometimes one has a prime knot, the $\triangle(t)$
of which can be written as a product

$$\triangle(t) = \triangle_1(t)\, \triangle_2(t) \ldots \triangle_n(t) \tag{6.1}$$

of $n$ polynomials some of which happen to be Alexander polynomials of
real knots! Hence, prime knots do not necessarily have prime Alexander
polynomials.

A third drawback of the Alexander polynomial occurs for stereoisomers. Two knots are called stereoisomers of each other if one can be obtained from the other by reflections in a plane. In the case of non-prime knots one can also take the mirror image(s) of some of the constituents knots. Stereoisomers always have the same $\triangle(t)$. Examples are the "granny"

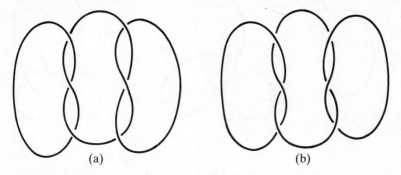

(a)                                        (b)

Fig. 4.6.   The granny knot (a) and the square knot (b) have the same Alexander polynomial.

knot and the "square" knot, drawn in Figs. 4.6a, b, both of which have $\triangle(t) = (t^2 - t + 1)^2$. So stereoisomers cannot be distinguished by their Alexander polynomials.

**4.7   Topological invariants for links**

A link consists of two or more knots which cannot be separated by continuous deformations without (self) intersections. The classification of links, just like the clasification of knots, first divides them according to the minimum number of double points of their projection on a plane.

For some fully reduced links one of the constitutent knots of the link can contain a loop which shows double points with itself. We shall call this a composite link, in contrast with a simple link which does not show such loops (examples are drawn in Fig. 4.7a, b, c). For example, all links of trivial knots are simple. In the existing tables of links [23, 31] only the simple links are listed according to increasing total number of double points. In the rest of this chapter we restrict outselves to links consisting of two knots.

*4.7.1.   The Gaussian invariant*

The earliest known invariant for two links was discovered more than 150 years ago by Gauss in a study of the mutual induction of two closed conducting wires $C$ and $D$ [37]. This is the linking number, winding number

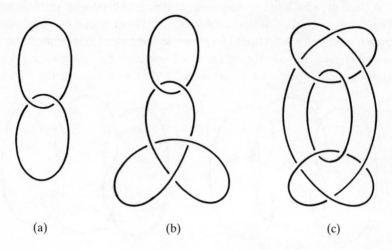

Fig. 4.7.   Two simple links (a,c) and one composite link (b).

or entanglement index of $C$ with respect to $D$, given by (1.10) or an equivalent form:

$$L[C, D] = (4\pi)^{-1} \oint_C d\mathbf{r} \cdot \oint_D |\mathbf{r} - \mathbf{s}|^{-3} \, d\mathbf{s} \times (\mathbf{r} - \mathbf{s}) \qquad (7.1)$$

$$= (4\pi)^{-1} \oint_C \oint_D |\mathbf{r} - \mathbf{s}|^{-3} \, (d\mathbf{r} \times d\mathbf{s}) \cdot (\mathbf{r} - \mathbf{s}) \ . \qquad (7.2)$$

The sign of $L$ depends only on the arbitrary orientation of $C$ and $D$; hence the relevant quantity is $|L|$.

Although $L$ is invariant under continuous non-intersecting deformations this integral is neither an appropriate topological invariant for links nor fit

Table 4.2.   The number ($M$) of simple links of two knots with $n$ double points, subdivided according to the absolute value of the winding number ($|L|$).

| $n$ | $M(n)$ | $|L|=0$ | 1 | 2 | 3 | 4 |
|---|---|---|---|---|---|---|
| 2 | 1 | 0 | 1 | 0 | 0 | 0 |
| 4 | 1 | 0 | 0 | 1 | 0 | 0 |
| 5 | 1 | 1 | 0 | 0 | 0 | 0 |
| 6 | 3 | 0 | 0 | 1 | 2 | 0 |
| 7 | 8 | 4 | 2 | 2 | 0 | 0 |
| 8 | 16 | 4 | 2 | 5 | 2 | 3 |
| 9 | 61 | 20 | 13 | 20 | 6 | 2 |
| Total: | 91 | 29 | 18 | 29 | 10 | 5 |

for the classification of knots. In order to demonstrate this we have listed in Table 4.2 the number $M(n)$ of simple links with $n$ double points. For each link we have calculated $|L|$; the subdivision of $M(n)$ according to the value of $|L|$ is listed in the remaining columns of the table. Now, it is easy to see that for a link with $n$ double points the maximum value of $L$ equals $n/2$. On the other hand, the table shows that $M(n)$ increases much faster than linearly with $n$. Hence there can be no one-to-one correspondence between the numerical values of $|L|$ and the different links. A serious consequence of this state of affairs is the case $L = 0$. Although the absence of linkage implies $L = 0$, the inverse does not hold: Table 4.2 shows that 29 out of the first 91 simple links have $L = 0$. Hence the Gaussian looping integral is of little use for the classification of links.

### 4.7.2. *The Alexander polynominal for links*

Alexander polynomials can be used for the classification of links in the same manner as for knots. Their construction proceeds in an analogous way. Moreover, the Alexander polynomials for knots and links are compatible with each other in a way to be discussed shortly. A complete description of the construction of the Alexander polynomial for a link can be found in [38–40]. The method can briefly be summarized in the following way, using a link of two knots $S$ and $T$ as an example. In the projection of the link on a plane one first follows $S$, numbering the underpasses in the order in which they are encountered. Next, one continues the numbering of the underpasses following the projection of $T$. The bridges are labeled with the numbers of the underpasses where they end. In the knot matrix, row $i$ corresponds to underpass $i$. In this row only 2 or 3 elements are non-zero: if underpass $i$ is part of $S$ and is bridged by a bridge of $S$ then this leads to 2 or 3 elements of the form $\pm 1$, $\pm s$ or $\pm(s-1)$. If underpass $i$ is part of $T$ and bridged by a bridge of $T$ one gets 2 or 3 non-vanishing elements of the form $\pm 1$, $\pm t$ or $\pm(t-1)$. If both underpass and bridge belong to different knots, only 3 elements of row $i$ are non-zero; they are of mixed form in $s$ or $t$. The Alexander polynomial $\Delta(s, t)$ of the link now equals the determinant of any $(n-1)\times(n-1)$ minor of the $n\times n$ knot matrix, after an appropriate normalization [The reader should be aware of an obvious misprint in the recipe of Ref. 40].

The Alexander polynomials for links have properties which are similar to those for knots. They are topological invariants. As a result of their palindromic symmetry they are specified by $m/2$ integers for an even value of the degree $m$ of $\Delta(s,t)$ and by $(m+1)/2$ integers for odd $m$. For all the explicitly analyzed links one has found $\Delta(s, t) \neq 0$. Moreover, the Alexander polynomial is connected with the Gauss invariant (7.1, 2) by the relation [41]

$$| L | = |\Delta(1, 1)| . \tag{7.3}$$

Finally, the compatibility of the Alexander polynomials of knots and links shows up in the property that $\Delta_c(s,t)$ for a composite link equals the product of the Alexander polynomial $\Delta_0(s,t)$ of the simple link and the Alexander polynomials $\Delta_1(s)$, $\Delta_2(t)$ of the two constitutent knots

$$\Delta_c(s, t) = \Delta_0(s, t)\Delta_1(s)\Delta_2(t) . \tag{7.4}$$

There are also some drawbacks in the use of these polynomials for links. Just as for knots there is no one-to-one correspondence between the polynomials and the links. For example, the 91 simplest links mentioned in Table 4.2 correspond to only 82 different Alexander polynomials.

Another drawback is the following. If two knots are not linked $\Delta(s,t) \equiv 0$ and hence

$$\Delta(-1, -1) = 0 . \tag{7.5}$$

The Russian authors [40] have used (7.5) as a criterion for the non-existence of a link. Unfortunately, for some complicated links (7.5) is true although the two constitutent knots are linked! This occurs, for example, in links $9_{54}^2$ and $9_{60}^2$ of Rolfsen's table [23]. This problem can easily be avoided by also substituting other values for $s$ and $t$, for example by also calculating $\Delta(1, 1)$.

A similar problem arises in connection with the remarkable statement in [40] that

$$|\Delta(-1, -1)| \geq 2 \tag{7.6}$$

for all links apart from the simplest one $2_1^2$. If one verifies this in Rolfsen's table one finds that $|\Delta(-1, -1)| = 1$ in the case of link $9_{49}^2$. Hence the *ad hoc* criterion (7.6) should be discarded.

## 4.8  Numerical hammagraphy

The numerical approach mentioned at the end of Section 4.4 was initiated by a group of Russian authors [38–40] and extended by des Cloizeaux and Mehta [42] and by Michels and Wiegel [32, 33]. In their pioneering study, the Russian group generated random walks on a cubic lattice. Des Cloizeaux and Mehta represented the chain by a Gaussian random walk in a continuous space. Strictly speaking the work reported in [38–40, 42] shows a flaw in the way in which the end points of the chain are joined. This is accomplished by making the one-step distribution function depend on the position along the backbone of the macromolecule in such a way that the end points are forced to coincide. This procedure, however, violates the independence of the different steps in the random walk. This flaw was later recognized and obviated by Chen [43] who, for the purpose of numerical hammagraphy, made rings by dimerization of two free random walks. In the present section we follow the method of Michels and Wiegel [32, 33] which uses molecular dynamics for an *a priori* closed ring model with repeating units of equal lengths.

Ring-shaped macromolecules are represented by sequences of $N$ points of equal mass $m$ at positions $r_1, r_2, \ldots, r_N$. Between any pair $(i, i+1)$ of nearest neighbors along the chain we assume a harmonic force around a fixed distance $l$:

$$\mathbf{F}_{i,i+1} = -k(|\,\mathbf{r}_{i+1} - \mathbf{r}_i\,| - l) \frac{\mathbf{r}_{i+1} - \mathbf{r}_i}{|\,\mathbf{r}_{i+1} - \mathbf{r}_i\,|} \qquad (8.1)$$

equals the force on mass $i$ due to mass $i+1$. Here $i = 1, 2, \ldots, N$ and $i+1 = 1$ if $i = N$. Moreover, a random force $\epsilon_i(t)$ acts on mass $i$, which models the heat motion of the surrounding solvent, and a frictional force $-\gamma d\mathbf{r}_i/dt$ models the hydrodynamic damping. The equations of motion are

$$m \frac{d^2\mathbf{r}_i}{dt^2} = \mathbf{F}_{i,i+1} - \mathbf{F}_{i-1,i} + \epsilon_i(t) - \gamma \frac{d\mathbf{r}_i}{dt}, \qquad (i = 1, 2, \ldots, N) \ . \qquad (8.2)$$

The values of $m$, $k$ and $\gamma$ are chosen in such a way that the dynamical modes of the system are strongly damped. Moreover, the value of $k$ is made large enough that at any time almost all (say 98%) of the repeating units have a momentary length near to (say within 2% of) the average length $l$.

At time $t = 0$ the molecule is assigned some initial configuration. One now solves the $N$ equations of motion (8.2) numerically with the algorithms commonly used in molecular dynamics (cf. Refs. 44–46). Once thermal equilibrium has been reached the configurations of the system on equally spaced times are stored in the computer's memory for further analysis. These time steps have to be chosen sufficiently large to avoid correlations between successive configurations. In this way microstates of a ring-shaped macromolecule are generated with the correct *a priori* probabilities. It should perhaps be stressed that the mass points, and the bonds between them, have no excluded volume. Hence the molecular dynamics carries the polymer freely through the various topological classes. For a further analysis of the geometry of the polymer generated by this procedure (usually called "Brownian dynamics") the reader might want to consult a recent note of Bishop and Michels [47].

### 4.8.1. *Probability of knots in a polymer ring*

The first basic question which numerical hammagraphy tries to answer is: which fraction of the stored configurations are knots and which fraction are topologically equivalent to a circle? In order to limit the computational task one does not use the full Alexander polynomial $\triangle(t)$, but one substitutes some specific value of $t$, which enables him to use as subroutines standard algorithms for the evaluation of determinants.

The whole numerical procedure consists of several steps. First, the stored configuration of the polymer is projected onto a plane. Second, most of the trivial loops are removed (numerically; remember: the computer cannot see!) in such a way that the number of double points decreases drastically. It turns out that this process reduces the number of double points considerably. For example, in a sample run with $N = 320$

Table 4.3. The fraction $\zeta$ of unknotted rings as a function of the chain length $(Nl)$ and the radius $(R)$ of the confining sphere in units $l$. In each case the total number of configurations studied is denoted by $t$, of which $b$ were found to be unknotted rings.

|         | $R/l$ | $t$ | $b$ | $\zeta$ |
|---------|-------|-------|-------|---------|
| $N = 64$ | $\infty$ | 25500 | 21548 | 0.845 |
|         | 4 | 17000 | 12247 | 0.720 |
|         | 3 | 16500 | 8972 | 0.544 |
|         | 2.5 | 16500 | 6239 | 0.378 |
|         | 2.1 | 19500 | 4207 | 0.216 |
|         | 1.9 | 14400 | 2014 | 0.140 |
| $N = 128$ | $\infty$ | 23500 | 15863 | 0.675 |
|         | 5 | 16500 | 6890 | 0.418 |
|         | 4 | 18100 | 4629 | 0.256 |
|         | 3.5 | 14400 | 2213 | 0.154 |
| $N = 192$ | $\infty$ | 38500 | 20789 | 0.540 |
|         | 8 | 21900 | 8782 | 0.401 |
|         | 6.3 | 18000 | 5099 | 0.283 |
|         | 5.5 | 16800 | 3443 | 0.205 |
|         | 5 | 14300 | 2166 | 0.151 |
| $N = 256$ | $\infty$ | 27420 | 15869 | 0.424 |
| $N = 320$ | $\infty$ | 44220 | 15156 | 0.343 |

Ref. 32 reports that this process removed all double points for about 85% of those configurations that turned out to be unknotted! Third, for those

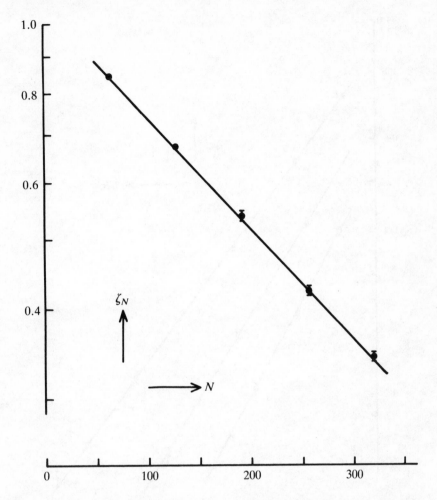

Fig. 4.8. Semi-logarithmic plot of the fraction $\zeta_N$ of unknotted rings as a function of the number of repeating units $N$. The uncertainty indicated by the error bars corresponds to $\pm 2\sigma_d$ where $\sigma_d$ denotes the standard deviation $[(1-\zeta_N)\zeta_N/t]^{1/2}$.

configurations which cannot be reduced any further one calculates the value of $\triangle(-1)$. Here one uses the empirical fact that a study of the existing knot tables shows that the relation

$$\triangle(-1) \neq \pm 1 \tag{8.3}$$

is a suitable criterion for the existence of a non-trivial knot.

The fraction $(\zeta_N)$ of unknotted configurations of a closed macromolecule consisting of $N$ repeating units is usually calculated for two geometries. In the first geometry, the molecule can move freely throughout all space. In

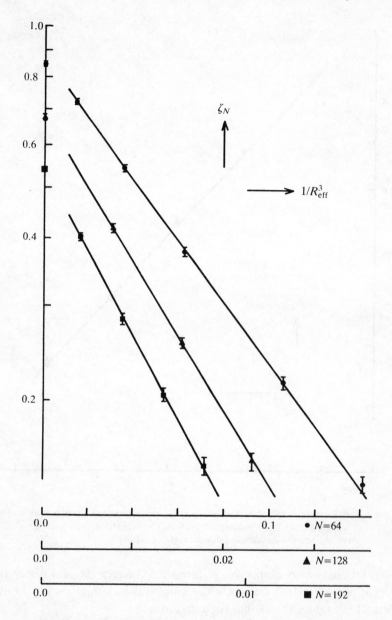

Fig. 4.9.   Semi-logarithmic plot of $\zeta_N$ as a function of $R_{\text{eff}}^{-3}$ for three different values of $N$. Error bars as in Fig. 4.8. Note that the wall of the sphere to the interior of which the macromolecule is confined, is represented by a harmonic force which acts as soon as the distance to the center of the sphere is larger than $R$. Because of the finiteness of this force the effective radius $R_{\text{eff}}$ of the sphere is slightly larger than $R$: in all cases but one we find $R_{\text{eff}} = R + 0.015l$. For the case $R/l = 1.9$ we find $R_{\text{eff}} = R + 0.02l$. In this case the radius of the sphere is comparable to the length of a repeating unit.

the second geometry, the molecule is confined to the interior of a sphere of radius $R$. The results are presented in Table 4.3. Here, $R$ is measured in units $l$. Of the total number of configurations ($t$) the number of unknotted configurations equals $b$; obviously $\zeta_N = b/t$. These results, which are taken from Ref. 32, 33, are also plotted in Figs. 4.8 and 4.9. When the results in Fig. 4.8 and Table 4.3 with $R/l = \infty$ are represented by a power law of the form

$$\zeta_N = C\mu^N N^\alpha , \qquad (8.4)$$

the best fit is found to be

$$\mu = 0.99640 \pm 0.00002 , \qquad (8.5)$$

$$\alpha = 0.0088 \pm 0.0006 , \qquad (8.6)$$

$$C = 1.026 \pm 0.003 , \qquad (8.7)$$

where the uncertainty indicated is given by $\pm$ twice the standard deviation. Thanks to the large number of configurations analyzed, one is able to determine the value of the "topological exponent" $\alpha$ with a fair numerical accuracy. It is remarkable that $\alpha$ turns out to be so close to zero.

The results in Fig. 4.9 and Table 4.3 for various values of $R/l$ and large $N$ can be represented by a scaling formula of the form

$$\zeta_N(R) = \exp[-A(N^\beta l/R)^\gamma] ; \qquad (8.8)$$

the best fit is found to be

$$\gamma = 3, \quad \beta = 0.76 . \qquad (8.9)$$

The scaling formulas (8.4) and (8.8) are reminiscent of similar formulas in the theory of critical phenomena. At the time of writing, the author can offer no explanation for the occurrence of scaling laws in these topological problems.

### 4.8.2 Hammagraphy of links: Gauss invariant versus Alexander polynomial

It was already pointed out in Subsection 4.7.1 that the Gauss invariant has some draw-backs as far as the classification of links is concerned. In view of the fact that this analytical invariant was used in Sections 4.1, 4.2, 4.3 to obtain exact solutions, one would like to compare the predictions of the Gauss invariant with those of the Alexander polynomial. In order to do this Michels and Wiegel [33] generated random polymer rings (of various lengths) with the molecular dynamics method outlined above. For each configuration they determined the center of gravity, and placed an infinite straight line through it. Next, the value of the Gauss invariant $|L|$ was calculated by counting the number of times that the polymer effectively encircled the straight line. The probabilities $P$ for the different values of

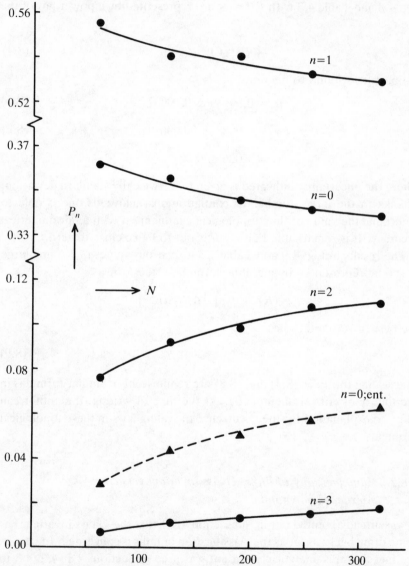

Fig. 4.10.   The probabilities for the different values of $n = |L|$ are plotted for different lengths $Nl$ of the polymer ring. The dashed line gives the probability of those cases in which $n = 0$ and $\triangle(s, t) \neq 0$.

$n = |L|$ are plotted in Fig. 4.10 for different lengths $Nl$ of the polymer. Next, we select those cases in which $n = 0$. For those cases, one closes the straight line at large distances from the polymer by an arbitrary polygon. After the reductions mentioned earlier in this section they calculated the value $\Delta(-1, -1)$ of the Alexander polynomial. Using the criterion (7.5) one determines how many of these polymer configurations as a fraction of

Table 4.4. The probability of finding the value $n$ for the absolute value of the winding number ($|L|$), as a function of the polymer length ($Nl$). The last column denotes the probability to find $n = 0$ *and* $\Delta(s, t) \neq 0$. For each case $t$ denotes the total number of configurations; each of them was sampled in 3 orthogonal projections.

| $N$ | $t$ | $n=0$ | $n=1$ | $n=2$ | $n=3$ | $n=4$ | $n=0$; ent |
|---|---|---|---|---|---|---|---|
| 64 | 40005 | 0.362 | 0.555 | 0.076 | 0.0066 | 0.0005 | 0.0286 |
| 128 | 41005 | 0.356 | 0.541 | 0.092 | 0.0109 | 0.0010 | 0.0435 |
| 192 | 45015 | 0.346 | 0.541 | 0.098 | 0.0131 | 0.0015 | 0.0504 |
| 256 | 48520 | 0.342 | 0.533 | 0.108 | 0.0153 | 0.0016 | 0.0573 |
| 320 | 49525 | 0.340 | 0.530 | 0.110 | 0.0179 | 0.0022 | 0.0633 |

the total $t$ are nevertheless entangled with the straight line. These results are also displayed in Table 4.4 and Fig. 4.10.

Inspection of the figure shows that the Gauss invariant leads to the erroneous conclusion that polymer and line are unlinked in about 8% of the cases with $N = 64$. This fraction of erroneous predictions increases up to 19% for $N = 320$. Note that even the longest polymers generated are short in the sense that a large fraction of their configurations is still unknotted (cf. Table 4.3). Nevertheless, already in this case the Gauss invariant underestimates the number of links significantly.

# References

[1] F.W. Wiegel, "Conformational Phase Transitions in a Macromolecule: Exactly Solvable Models," in *Phase Transitions and Critical Phenomena*, vol. 7., C. Domb and J.L. Lebowitz, eds. (Academic Press, New York, 1983) p. 101.

[2] F.W. Wiegel, *Physica* **109A** (1981) 609.

[3] F.B. Fuller, in "Mathematical Problems in the Biological Sciences," *Proc. Symp. Appl. Math.* **14** (1962) 64 (American Mathematical Society, Providence). Also compare *Proc. Nat. Acad. Sci.* USA **68** (1971) 815; **75** (1978) 3557; F.H.C. Crick, *Proc. Nat. Acad. Sci.* USA **73** (1976) 2639; C.J. Benham, *J. Mol. Biol.* **123** (1978) 361.

[4] J.D. Jackson, *Classical Electrodynamics* (Wiley, New York, 1975).

[5] S. Prager and H.L. Frisch, *J. Chem. Phys.* **46** (1967) 1475.

[6] S.F. Edwards, *Proc. Phys. Soc.* **91** (1967) 513. Also compare R. Alexander-Katz and S.F. Edwards, *J. Phys.* **A5** (1972) 674.

[7] N. Saito and Y. Chen, *J. Chem. Phys.* **59** (1973) 3701.

[8] F.W. Wiegel, *J. Chem. Phys.* **67** (1977) 469. Also compare A. Inomata and V.A. Singh, *J. Math. Phys.* **19** (1978) 2318; V. Tanikella and A. Inomata, *Phys. Lett.* **87A** (1982) 196.

[9] M. Abramowitz and I. A. Stegun, *Handbook of Mathematical Functions* (Dover, New York, 1970).

[10] F.W. Wiegel and J. Boersma, *Physica* **122A** (1983) 325.

[11] J. Boersma and F.W. Wiegel, *Physica* **122A** (1983) 334.

[12] F.W. Wiegel, *J. Math. Phys.* **21** (1980) 2111; *Fluid Flow through Porous Macromolecular Systems,* Lecture Notes in Physics 121 (Springer, Berlin, 1980) p. 78.

[13] H.S. Carslaw, *Proc. London Math. Soc.* **(1) 30** (1899) 121.

[14] H.S. Carslaw and J.C. Jaeger, *Conduction of Heat in Solids* (Oxford University Press, Oxford, 1959) Sec. 14.14 III.

[15] L.S. Schulman, in *Proceedings of the Symposium on Wave-Particle Dualism,* Perugia (1982) to be published.

[16] L.S. Schulman, *Phys. Rev. Lett.* **49** (1982) 599.

[17] S.W. Lee, *J. Math. Phys.* **19** (1978) 1414.

[18] C.F. Klein, *Ann. N.Y. Acad. Sci.* **385** (1982) 1.

[19] W.J. Conklin, *Ann. N.Y. Acad. Sci.* **385** (1982) 261.

[20] R.H. Fox, "A Quick Trip through Knot Theory," in *Topology of 3-manifolds,* M.K. Fort, ed. (Prentice-Hall, Engelwood Cliffs, 1962) p. 120.

[21] K. Reidemeister, *Knotentheorie* (Springer Verlag, Heidelberg, 1974).

[22] R.H. Crowell and R.H. Fox, *Introduction to Knot Theory* (Springer Verlag, Heidelberg, 1977). Also.compare B. Duplantier, *Commun. Math. Phys.* **82** (1981) 41; **85** (1982) 221; *C.R. Acad. Sci.* Paris **293**, Série I (1981) 693.

[23] D. Rolfsen, *Knots and Links* (Publish or Perish Inc., Berkeley, 1976).

[24] M. Delbrück, in "Mathematical Problems in the Biological Sciences," *Proc. Symp. Appl. Math.* **14** (1962) 55 (American Mathematical Society, Providence). Also compare G.M. Crippen, *J. Theor. Biol.* **45** (1974) 327; M. LeBret, *Biopol* **19** (1980) 619.

[25] D.S. McKenzie, *Phys. Reports* **27** (1976) 35.

[26] P.G. de Gennes, *Scaling Concepts in Polymer Physics* (Cornell University Press, Ithaca, 1979).

[27] C.N. Little, *Trans. Conn. Acad. Sci.* **18** (1885) 374.

[28] C.N. Little, *Trans. Roy. Soc.* Edin. **36** (1890) 253.

[29] C.N. Little, *Trans. Roy. Soc.* Edin. **39** (1900) 771.

[30] J.W. Alexander and G.B. Briggs, *Ann. Math.* **28** (1927) 562.

[31] J.H. Conway, in *Computational Problems in Abstract Algebra* (Pergamon Press, Oxford, 1969) p. 329.

[32] J. P.J. Michels and F.W. Wiegel, *Phys. Lett.* **90A** (1982) 381.

[33] J.P.J. Michels and F.W. Wiegel, *Proc. Roy. Soc.* London **A403** (1986) 269.

[34] S. Kinoshita and H. Terasaka, *Osaka Math. J.* **9** (1957) 131.

[35] W. Magnus and A. Peluso, *Comm. Pure Appl. Math.* **20** (1967) 749.

[36] Quoted in R. Riley, *Math. of Comp.* **25** (1971) 603.

[37] C.F. Gauss, *Koenig Ges. Wiss.* Goettingen **5** (1877) 602.

[38] A.V. Vologodskii, A.V. Lukashin, M.D. Frank-Kamenetskii and V.V. Anshelevich, *Sov. Phys. JETP* **39** (1974) 1059.

[39]  M.D. Frank-Kamenetskii, A.V. Lukashin and A.V. Vologodskii, *Nature* **258** (1975) 398.

[40]  A.V. Vologodskii, A.V. Lukashin and M.D. Frank-Kamenetskii, *Sov. Phys. JETP* **40** (1975) 932.

[41]  R. Ball and M.L. Mehta, *J. de Phys.* **42** (1981) 1193.

[42]  J. des Cloizeaux and M.L. Mehta, *J. de Phys.* **40** (1979) 665.

[43]  Y. Chen, *J. Chem. Phys.* **75** (1981) 2447.

[44]  L. Verlet, *Phys. Rev.* **159** (1967) 98.

[45]  J.P.J. Michels, *Physica* **90A** (1978) 179.

[46]  A.J. Kox, J.P.J. Michels and F.W. Wiegel, *Nature* **287** (1980) 317.

[47]  M. Bishop and J.P.J. Michels, *J. Chem. Phys.* **82** (1985) 1059.

# V.  QUANTUM PHYSICS

### 5.1  Advantages of the Feynman path integral

After the more technical calculations in the last three chapters we want to slow down for a short while and give a bird's eye view of the applications of path integration to quantum physics in a simple qualitative way. The idea is as follows. Take a classical system. The physical state of this system is characterized by a set of variables which can be functions of time; this will be called the "path" of the system. Almost all classical systems have a Lagrangian $L$ which is a functional $L[\text{path}]$ of the path of the system. The action $S$ is defined as the time integral of this Lagrangian

$$S[\text{path}] = \int_{t_a}^{t_b} L[\text{path}]\, dt \ . \tag{1.1}$$

The main requirement to be imposed on the form of the Lagrangian is that the variational principle

$$\delta S = 0 \tag{1.2}$$

leads to the classical equations of motion for the system. Now, in quantum physics the time evolution of this system is characterized by a propagator $G(x_b, t_b \mid x_a, t_a)$, where $x_a$ denotes the state of the system at time $t_a$, and $x_b$ the state of the system at some later time $t_b$. The physical significance of the propagator is contained in the formula

$$\psi(b) = \int G(b \mid a) \, \psi(a) \, dx_a \, , \tag{1.3}$$

where $\psi$ denotes the wave function and where the integration is performed over all variables which characterize the state of the system at time $t_a$. It will be shown shortly that the propagator can be represented by the Feynman path integral

$$G(x_b, t_b \mid x_a, t_a) = \int_{x_a, t_a}^{x_b, t_b} \exp\left\{\frac{i}{\hbar} S[\text{path}]\right\} d[\text{path}] \, , \tag{1.4}$$

where $\hbar$ denotes Planck's constant divided by $2\pi$. It is for this reason that almost every problem in theoretical physics can be formulated in the language of path integration.

What are the advantages of formulating quantum physics starting with the last equation rather than solving the propagator from the Hamiltonian $H$ by way of the Schrödinger equation

$$i\hbar \frac{\partial G}{\partial t_b} = HG \tag{1.5}$$

(where all operators in $H$ act on the variables $t_b$, $x_b$)? The answer to this question is determined by one's personal perspective of physics, but the following points should be mentioned: (a) For quantization of a system with the path integral one needs the Lagrangian instead of the Hamiltonian. This language is therefore especially appropriate when the classical equations of motion of the system can be cast in the Lagrangian—but not in the Hamiltonian form. Such is the case for systems which show dissipation, but also for the general theory of relativity where the "path" of the system in space-time is defined as a set of 10 components $g_{\mu\nu}$ ($x^1, x^2, x^3, x^4$) of the metric tensor $g$ as functions of the four space-time coordinates $x^1$, $x^2, x^3, x^4$. In this way Hawking has quantized the metric field of a black hole [1–7]. (b) In many cases the path-integral representation shows immediately which paths of the system are important. For example, the semiclassical region of quantum physics is characterized by the order-of-magnitude estimate

$$S \gg \hbar \tag{1.6}$$

for the classical action, i.e., the action calculated with the classical path. In this case the contributions of most paths to the integral (1.4) will cancel each other, unless these paths are somehow "close" to the solution of $\delta S = 0$, which is the classical path. In the semiclassical region the propagator will therefore be dominated by those paths which are in the immediate vicinity of the classical path; the size of this vicinity follows from the estimate $\delta S \sim \hbar$. (c) In this way one finds new approximation methods which generalize the semiclassical WKB method. For example, we may write

$$\text{path} = \text{classical path} + \text{fluctuation} , \qquad (1.7)$$

expand $S$ up to the second order in the fluctuation and perform the integration over the fluctuations with the methods of Chapters I and II. This leads to an approximation of the form

$$G \cong \sum \exp\left\{\frac{i}{\hbar} S[\text{classical path}]\right\} N[\text{classical path}] , \qquad (1.8)$$

where the factor $N$ arises from the path integral over the fluctuations and where the sum extends over all solutions of the classical equations of motion which are consistent with the boundary conditions. These classical solutions have different names in different parts of theoretical physics: "metastable droplets" in the theory of phase transitions, "quantized vortex lines" in the theory of the Bose fluid, "quantized flux lines" in the theory of superconductors, "instantons" in quantum field theory, etc. (d) These new approximation methods have a non-perturbative character and often lead to non-analytic results. (e) With the use of path integrals one can sometimes demonstrate a one-to-one correspondence between problems which derive from altogether different parts of physics. This enables one to "borrow" exact solutions and various approximation schemes. An example is the relation between the Aharonov-Bohm effect in quantum physics and some of the polymer entanglement problems discussed in Chapter IV.

## 5.2 The Feynman path integral and the propagator for the Schrödinger equation

The simplest way to derive the Feynman path-integral expression (1.4) for a particle in an external potential $V(x)$ is to start with the usual formulation of quantum mechanics, which forms the subject of various monographs like the excellent textbook by Bohm [8], and which is based upon the Schrödinger equation. Now, compare this equation

$$i\hbar \frac{\partial G}{\partial t} = -\frac{\hbar^2}{2m}\frac{\partial^2 G}{\partial x^2} + V(x)G \ , \tag{2.1}$$

with the equation (I.4.8) for the propagator of a Brownian particle in an absorbing medium

$$\frac{\partial G_A}{\partial \tau} = D\frac{\partial^2 G_A}{\partial \xi^2} - A(\xi)G_A \ . \tag{2.2}$$

This suggests the correspondence

$$G <=> G_A \ , \qquad V <=> A \ , \qquad \hbar^2/2m <=> D \ ,$$

$$it/\hbar <=> \tau \ , \qquad x <=> \xi \ . \tag{2.3}$$

If these substitutions are made in (I.4.5) and (I.2.11) one finds

$$G(x, t \mid x_0, t_0) = \int_{x_0, t_0}^{x, t} \exp\left\{\frac{i}{\hbar}\int_{t_0}^{t}\left[\frac{1}{2}m\left(\frac{dx}{dt'}\right)^2 - V(x)\right]dt'\right\}$$

$$\times d[x(t')] \ , \quad (t > t_0) \tag{2.4}$$

where the path-integral symbol indicates the multiple integral

$$\int d[x(t')] <=> (2\pi i\hbar\varepsilon/m)^{-(N+1)/2}\int_{-\infty}^{+\infty}dx_1\int_{-\infty}^{+\infty}dx_2\ldots\int_{-\infty}^{+\infty}dx_N \tag{2.5}$$

in the limit $\varepsilon \to 0$, $N \to \infty$, $(N+1)\varepsilon = t - t_0$. In this way one is lead to the Feynman path integral (1.4).

Because of the importance of the result (2.4, 5) it is of interest to find a derivation of it which is independent of the correspondence (2.3). This can be done, for example, following the method of Section 1.5. Another method is the following. Perform all integrations over $x_1, x_2 \ldots x_{N-1}$ in (2.5), keeping $x_N$ fixed. Taking the limit $\varepsilon \to 0$ for all intervals $(t_0, t_1)$, $(t_1, t_2), \ldots, (t_{N-1}, t_N)$ but not for $(t_N, t)$ one finds that the function $G(x, t \mid x_0, t_0)$ is the solution of the integral equation

$$G(x, t \mid x_0, t_0) = \lim_{\varepsilon \to 0}(2\pi i\hbar\varepsilon/m)^{-1/2}\int_{-\infty}^{+\infty}G(x_N, t_N \mid x_0, t_0) \tag{2.6}$$

$$\times \exp\left\{\frac{i}{\hbar}\varepsilon\left[\frac{1}{2}m\left(\frac{x - x_N}{\varepsilon}\right)^2 - V(x_N)\right]\right\}dx_N \ .$$

The first term in the exponential shows that the values of $x_N$ which contribute appreciably to this integral have $x_N = x + O(\varepsilon^{1/2})$. Hence we write, suppressing the argument $x_0$, $t_0$,

$$G(x_N, t_N) = G(x, t_N) + (x_N - x) \frac{\partial}{\partial x} G(x, t_N)$$

$$+ \frac{1}{2} (x_N - x)^2 \frac{\partial^2}{\partial x^2} G(x, t_N) + O(x_N - x)^3 , \tag{2.7}$$

$$\exp\left\{ -\frac{i\varepsilon}{\hbar} V(x_N) \right\} = 1 - \frac{i\varepsilon}{\hbar} V(x_N) + O(\varepsilon^2) . \tag{2.8}$$

Combination of the last three equations leads to the unelegant expression

$$G(x, t) = G(x, t_N) (2\pi i\hbar\varepsilon/m)^{-1/2} \int_{-\infty}^{\infty} \exp\left\{ \frac{im}{2\hbar\varepsilon} (x - x_N)^2 \right\} dx_N$$

$$+ \frac{1}{2} \frac{\partial^2}{\partial x^2} G(x, t_N) (2\pi i\hbar\varepsilon/m)^{-1/2} \int_{-\infty}^{\infty} (x - x_N)^2$$

$$\times \exp\left\{ \frac{im}{2\hbar\varepsilon} (x - x_N)^2 \right\} dx_N$$

$$- \frac{i\varepsilon}{\hbar} G(x, t_N) (2\pi i\hbar\varepsilon/m)^{-1/2} \int_{-\infty}^{+\infty} V(x_N)$$

$$\times \exp\left\{ \frac{im}{2\hbar\varepsilon} (x - x_N)^2 \right\} dx_N + O(\varepsilon^{3/2}) , \tag{2.9}$$

in the limit $\varepsilon \to 0$. The three integrals with their normalization factors equal 1, $i\hbar\varepsilon/m$ and $V(x)$, respectively; so

$$G(x, t) - G(x, t_N) = \frac{i\hbar\varepsilon}{2m} \frac{\partial^2}{\partial x^2} G(x, t_N) - \frac{i\varepsilon}{\hbar} V(x) G(x, t_N) + O(\varepsilon^{3/2}) . \tag{2.10}$$

If both sides of this equation are multiplied with $i\hbar/\varepsilon$ and the limit $\varepsilon \to 0$ is taken one finds

$$i\hbar \frac{\partial G}{\partial t} = -\frac{\hbar^2}{2m} \frac{\partial^2 G}{\partial x^2} + VG , \tag{2.11}$$

which shows that $G$ is a solution of the Schrödinger equation. It is also very easy to show that

$$\lim_{t \to t_0} G(x, t \mid x_0, t_0) = \lim_{\varepsilon \to 0} (2\pi i \hbar \varepsilon/m)^{-1/2}$$

$$\times \exp\left\{\frac{im}{2\hbar\varepsilon}(x - x_0)^2\right\} = \delta(x - x_0) \ . \quad (2.12)$$

Hence the function defined by Eqs. (2.4,5) is the propagator of the Schrödinger equation.

In view of the existence of many textbooks about quantum mechanics, especially the book by Feynman and Hibbs, there is no need to pursue this topic in detail in the present monograph. We merely use the correspondence (2.3) to derive a list of explicit expressions for the propagator of a quantum mechanical particle in a variety of external potentials.

*A free particle in one dimension*: In this case $V(x) = 0$, hence Eq. (2.3) tells us that $A(x) = 0$ in the corresponding diffusion problem. The propagator was given by Eq. I.2.5; upon "translation" one finds for the propagator of the Schrödinger equation

$$G_0(x, t \mid x_0, t_0) = \{2\pi i \hbar(t - t_0)/m\}^{-1/2} \exp\left\{+i\,\frac{m(x - x_0)^2}{2\hbar(t - t_0)}\right\} \quad (2.13)$$

for $t \geq t_0$. Of course $G_0$ is defined to equal 0 for $t < t_0$.

*A free particle in d dimensions:* This result can, of course, be immediately generalized to the motion of a free quantum mechanical particle in a $d$-dimensional space. One finds

$$G_0(\mathbf{r}, t \mid \mathbf{r}_0, t_0) = \{2\pi i \hbar(t - t_0)/m\}^{-d/2} \exp\left\{+i\,\frac{m \mid \mathbf{r} - \mathbf{r}_0 \mid^2}{2\hbar(t - t_0)}\right\} \quad (2.14)$$

for $t \geq t_0$.

*A particle in a harmonic potential*: If a particle moves along a line (the $x$-axis) under the influence of a harmonic force, the potential is given by

$$V(x) = \frac{1}{2} m\omega^2 x^2 \ , \quad (2.15)$$

where $\omega$ is the angular velocity of the classical motion. Equation (2.3) tells us that $A(x) = \frac{1}{2}m\omega^2 x^2$ in the corresponding diffusion problem. The propagator was calculated in Section 1.6. Translating Eq. (I.6.15) back with the help of (2.3) one finds

$$G(x, t \mid x_0, t_0) = \left\{\frac{2\pi i \hbar}{m\omega} \sin\omega(t - t_0)\right\}^{-1/2} \quad (2.16)$$

$$\times \exp\left\{+\frac{im\omega}{2\hbar}\,\frac{(x^2 + x_0^2)\cos\omega(t - t_0) - 2xx_0}{\sin\omega(t - t_0)}\right\}$$

for $t \geq t_0$. If the particle moves in a $d$-dimensional space under the influence of an anisotropic harmonic potential of the form

$$V(x_1, x_2, \ldots, x_d) = \frac{1}{2}m \sum_{j=1}^{d} \omega_j^2 x_j^2 \tag{2.17}$$

the propagator will of course be given by a product of $d$ factors, each of which has the form of (2.16) with the appropriate classical frequency $\omega_1$, $\omega_2, \ldots, \omega_d$.

*A particle in a plane with a half-line barrier*: In Section 4.3 we derived the propagator for a particle which moves in a plane (cf. Fig. 4.1) in which there is a hard wall which has the shape of a half-line located, in polar coordinates, at $\theta = \pm \pi$, $0 < r < \infty$. On this wall the boundary condition is $G = 0$, which implies Dirichlet boundary conditions in the corresponding diffusion problem. Equation (IV. 3.25) gives for the propagator between points with polar coordinates $(r_0, \theta_0)$ and $(r, \theta)$

$$G(r, \theta, t \mid r_0, \theta_0, 0) = (2\pi i\hbar t/m)^{-1} \exp\left(\frac{imR^2}{2\hbar t}\right) F\left(\sqrt{\frac{2mr_0 r}{\hbar t}} \cos\frac{1}{2}(\theta - \theta_0)\right)$$

$$- (2\pi i\hbar t/m)^{-1} \exp\left(\frac{imR_1^2}{2\hbar t}\right) F\left(-\sqrt{\frac{2mr_0 r}{\hbar t}} \cos\frac{1}{2}(\theta + \theta_0)\right), \tag{2.18}$$

where $F$ denotes the Fresnel integral

$$F(z) \equiv \frac{1}{\sqrt{i\pi}} \int_{-\infty}^{z} e^{i\xi^2} d\xi, \tag{2.19}$$

and where

$$R^2 = r^2 + r_0^2 - 2rr_0 \cos(\theta - \theta_0), \tag{2.20}$$

$$R_1^2 = r^2 + r_0^2 - 2rr_0 \cos(\theta + \theta_0). \tag{2.21}$$

For the corresponding three-dimensional problem the particle moves in space in the presence of a hard wall which has the shape of a half-plane located, in Cartesian coordinates, at $-\infty < x < 0$, $y = 0$, $-\infty < z < +\infty$. The propagator will be given by the product of a factor similar to (2.18) for the projection of the movement onto the $x$, $y$ plane and a factor similar to (2.13) for the projection of the movement along the $z$-axis.

The propagators listed in this section (and those for an electron in a constant electric and/or magnetic field, which are simple variations on the result (2.16), cf. [I.34]) are the only cases for which the propagator of the Schrödinger equation is known explicitly.

## 5.3 The Aharonov-Bohm effect

The Aharonov-Bohm effect demonstrates that a charged particle which passes through a region of space in which both the electric field **E** and the magnetic field **H** vanish can still be influenced by the vector potential. In such a region the total electromagnetic force

$$\mathbf{F} = q\mathbf{E} + \frac{q}{c}\,\mathbf{v} \times \mathbf{H} \tag{3.1}$$

vanishes (here $q$ is the charge and $c$ the speed of light). Yet, according to quantum mechanics, the motion of the particle is described by the Feynman path integral for the propagator (1.4) in which the action equals the time integral of the Lagrangian (cf. Ref. 9).

$$L = \frac{1}{2}m\left(\frac{d\mathbf{r}}{dt}\right)^2 + \frac{q}{c}\,\mathbf{A} \cdot \frac{d\mathbf{r}}{dt} - \phi - V \;. \tag{3.2}$$

In this expression **A** equals the vector potential, $\phi$ the electric potential and $V$ the potential of some other external, non-electromagnetic, force field. The Aharonov-Bohm effect is due to the fact that if **E** and **H** vanish in some region $R$ of space, **A** need not vanish in $R$ due to the presence of magnetic fields outside $R$.

The literature on this surprising effect has been reviewed by Kobe [10]. In this section we consider the Aharonov-Bohm effect due to a magnetic flux $\Phi$ confined to a solenoid. The solenoid will be assumed to be infinitesimally thin and in the form of a closed continuous curve $D$. For this magnetic field the vector potential is given by

$$\mathbf{A}(\mathbf{r}) = \frac{\Phi}{4\pi} \oint_D |\,\mathbf{r} - \mathbf{s}\,|^{-3}\, d\mathbf{s} \times (\mathbf{r} - \mathbf{s}) \;. \tag{3.3}$$

where $d\mathbf{s}$ denotes the line element of $D$. The electric potential $\phi$ equals a constant, which can be taken as the zero of energy.

Now consider an arbitrary path $\mathbf{r}(t)$ in the integral (1.4). Substituting the Lagrangian (3.2) into the action (1.1) the term with the magnetic vector potential leads to a term in the exponential

$$\frac{iq}{\hbar c}\int_{t_a}^{t_b} \mathbf{A} \cdot \frac{d\mathbf{r}}{dt}\, dt = \frac{iq}{\hbar c}\int_C \mathbf{A} \cdot d\mathbf{r} \;, \tag{3.4}$$

where the line integral extends along the contour $C$ traced out by the charged particle in space. This line integral has a simple property which follows immediately if one notes that $\mathbf{A}$ as defined by (3.3) equals the magnetic flux $\Phi$ times the vector field $\mathbf{F}$ defined by Eq. (IV.1.6) in the context of polymer entanglements:

$$\int_C \mathbf{A} \cdot d\mathbf{r} = \int_{C_0} \mathbf{A} \cdot d\mathbf{r} + n\Phi . \tag{3.5}$$

Here $C_0$ denotes some standard contour—for example, some continuous curve which connects $\mathbf{r}_a$ with $\mathbf{r}_b$ and which does not pass through $D$—and the integer $n$ equals the number of times that the closed curve $C - C_0$ winds through $D$ (see Fig. 5.1).

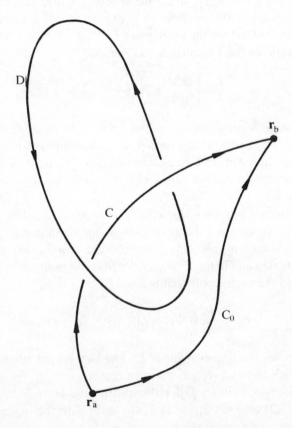

Fig. 5.1.   Magnetic flux tube $D$, with an arbitrary trajectory $C$ of a charged particle moving from $\mathbf{r}_a$ to $\mathbf{r}_b$, and some fixed standard trajectory $C_0$ which does not pass through $D$. With $-C_0$ one means the trajectory $C_0$ traversed in the opposite direction.

Substitution of the last two equations and (3.2) into (1.4) gives the following expression for the propagator

$$G(\mathbf{r}_b, t_b \mid \mathbf{r}_a, t_a) = \exp\left(\frac{iq}{\hbar c} \int_{C_0} \mathbf{A} \cdot d\mathbf{r}\right) \sum_{n=-\infty}^{+\infty} F_n(\mathbf{r}_b, t_b \mid \mathbf{r}_a, t_a)$$

$$\times \exp\left(\frac{iq\Phi}{\hbar c} n\right) , \tag{3.6}$$

where $F_n$ is the topologically constrained Feynman path integral

$$F_n(\mathbf{r}_b, t_b \mid \mathbf{r}_a, t_a) = \int_{\mathbf{r}_a, t_a}^{\mathbf{r}_b, t_b} \exp\left\{\frac{i}{\hbar} \int_{t_a}^{t_b} \left[\frac{1}{2}m\left(\frac{d\mathbf{r}}{dt}\right)^2 - V(\mathbf{r})\right] dt\right\} d_n[\mathbf{r}(t)] ,$$
$$\tag{3.7}$$

which has to be extended over all trajectories which (when completed into a closed loop by continuing them with $-C_0$) have an entanglement index $n$ with the magnetic flux line $D$. Hence $F_n$ is the quantum analog of the topologically constrained Wiener integrals which formed the subject of Chapter IV.

The result (3.6, 7) generalizes some ideas formulated more than a decade ago by Schulman [11]. It shows that for a solenoid of arbitrary shape, with arbitrary external forces, no Aharonov-Bohm effect will occur if the magnetic flux obeys the quantization condition

$$\frac{q\Phi}{hc} = 0, \pm 1, \pm 2, \dots \tag{3.8}$$

In this case all phase factors in (3.6) equal unity, and the summation over $n$ gives

$$G(\mathbf{r}_b, t_b \mid \mathbf{r}_a, t_a) = \exp\left(\frac{iq}{\hbar c} \int_{C_0} \mathbf{A} \cdot d\mathbf{r}\right) G_V(\mathbf{r}_b, t_b \mid \mathbf{r}_a, t_a) , \tag{3.9}$$

where $G_V$ denotes the full, unconstrained, propagator for a particle in a scalar potential $V$, in the absence of the magnetic vector potential. If the magnetic flux does not obey the quantization condition (3.8) the contributions of the various topological classes to the propagator (3.6) will interfere, and when a screen is placed behind the flux line the interference pattern on the screen will shift when $\Phi$ is increased. This is the Aharonov-Bohm effect.

## 5.4 Calculational techniques for constrained Feynman path integrals

In this final section we collect the various recipes for the calculation of Feynman integrals over paths with topological constraints, taking the integral (3.7) of the Aharonov-Bohm effect as an example. We follow Ref. 12.

(a) The first method has been discussed fully in Section 4.1. The idea is to find a functional $L[\mathbf{r}(t)]$ of the trajectory of the particle with the properties (IV.1.2). With this functional one can transform the constrained path integral (3.7) into an unconstrained one (the quantum analog of Eq. (IV.1.5)) which is subsequently calculated with a differential equation.

(b) Of course, one can try to calculate the full particle propagator (3.6) directly from the Schrödinger equation by solving the eigenvalue problem

$$H\psi_k = E_k\psi_k \ , \tag{4.1}$$

$$H = \frac{1}{2m}\left(\frac{\hbar}{i}\,\nabla - \frac{q}{c}\,\mathbf{A}\right)^2 + V \ , \tag{4.2}$$

and expanding in the orthonormal eigenfunctions

$$G(\mathbf{r}_b, t_b \mid \mathbf{r}_a, t_a) = \sum_k \psi_k(\mathbf{r}_b)\,\psi_k^*(\mathbf{r}_a)\,\exp\left\{\frac{E_k}{i\hbar}\,(t_b - t_a)\right\} \ . \tag{4.3}$$

If one succeeds in evaluating this sum explicitly the result can be expanded in integer powers of the phase factor $\exp(iq\Phi/\hbar c)$. A glance at (3.6) shows that the coefficient of the $n$th power of this factor equals the Feynman integral over the paths which are $n$ times entangled with the flux line, times a trivial phase factor $\exp\{(iq/\hbar c)\int_{C_0} \mathbf{A} \cdot d\mathbf{r})\}$. A calculation of this type is suggested in footnote 1 of a paper by Bernido and Inomata [13].

(c) Another calculational technique is the following. Let $S$ denote a surface of which $D$ is the boundary and which is oriented in such a way that the outward normal vector is related to the direction of $D$ by Maxwell's right-hand rule. Let the symbol $\mathbf{r} \uparrow S$ denote a limiting process in which the point $\mathbf{r}$ approaches some point of $S$ moving in the same direction as the outward normal vector. Similarly, $\mathbf{r} \downarrow S$ denotes the limit in which $\mathbf{r}$ approaches the same point of $S$ moving in the opposite direction. It follows from these considerations that the constrained path integrals $F_n(\mathbf{r}, t \mid \mathbf{r}_a, t_a)$ defined similar to (3.7), but with $n$ counting the number of times that the particle's trajectory pierces the surface $S$, are connected on $S$ by the boundary conditions

$$\lim_{r \uparrow S} F_n(\mathbf{r}, t | \mathbf{r}_a, t_a) = \lim_{r \downarrow S} F_{n+1}(\mathbf{r}, t | \mathbf{r}_a, t_a) \cdot$$
$$(\mathbf{r} \in S; n = 0, \pm 1, \pm 2, \ldots) \cdot \tag{4.4}$$

Similar relations hold for all partial derivatives. As all functions $F_n$ should be solutions of the same Schrödinger equation

$$i\hbar \frac{\partial F_n}{\partial t} = \left( -\frac{\hbar^2}{2m} \Delta + V(\mathbf{r}) \right) F_n \tag{4.5}$$

the problem is reduced to solving this simpler equation (which does not contain the vector potential!) but with the unusual boundary conditions (4.4) on $S$. This method has been used in [14] and [III.6].

(d)   For the case in which the flux tube is an infinite straight line one can also use a method due to Inomata and Singh [15] in which the constrained path integral is evaluated directly in polar coordinates.

## References

[1]   C.W. Misner, *Rev. Mod. Phys.* **29** (1957) 497.

[2]   J.R. Klauder, ed., *Magic Without Magic* (W.H. Freeman, San Francisco, 1972).

[3]   R.P. Feynman, in *Magic without Magic,* J.R. Klauder, ed. (W.H. Freeman, San Francisco, 1972) pp. 355 and 377.

[4]   J.V. Narlikar and T. Padmanabhan, *Phys. Reports* **100** (1983) 151.

[5]   G.W. Gibbons and S.W. Hawking, *Phys. Rev.* **D15** (1977) 2752.

[6]   S.W. Hawking and W. Israel, eds., *General Relativity* (Cambridge University Press, Cambridge, 1979).

[7]   S.W. Hawking in *General Relativity* , S.W. Hawking and W. Israel, eds. (Cambridge University Press, Cambridge, 1979) p. 746.

[8]   D. Bohm, *Quantum Theory* (Prentice-Hall, Englewood Cliffs, 1951).

[9]   H. Goldstein, *Classical Mechanics* (Addison-Wesley, Reading, 1959) p. 21.

[10]   D.H. Kobe, *Ann. Phys.* (N.Y.) **123** (1979) 381.

[11]   L.S. Schulman, *J. Math. Phys.* **12** (1971) 304.

[12]   F.W. Wiegel, *Physica* **109A** (1981) 609.

[13]   C. Bernido and A. Inomata, *Phys. Lett.* **77A** (1980) 394.

[14]   F.W. Wiegel, *J. Chem. Phys.* **67** (1977) 469.

[15]   A. Inomata and V.A. Singh, *J. Math. Phys.* **19** (1978) 2318.

# VI. CLASSICAL STATISTICAL PHYSICS

The partition function of a classical many-body system can almost always be represented by an integral over function space. The first section of this chapter introduces the concept of an integral over Gaussian random functions, which is a generalization of the Wiener integral. In the second section we show how the classical partition function can be written as a functional integral over these random functions. The last section treats an exactly solvable model for the condensation of a one-dimensional gas with this method.

## 6.1 Gaussian random functions

Consider a collection of functions $\phi(x)$ defined in a domain $\Omega$ of a finite dimensional Cartesian space. Given a functional $F[\phi(x)]$ of these functions we introduce the average $\langle F[\phi(x)] \rangle$ over the collection of functions. The explicit definition of the averaging process will be postponed for a while, but for any definition of the average one can introduce the characteristic functional

$$G[\xi(x)] = \left\langle \exp\{i \int_\Omega \xi(x) \, \phi(x) \, dx\} \right\rangle , \qquad (1.1)$$

which is defined for functions $\xi(x)$. If the characteristic functional is known, the $n$-point correlation function $\langle \phi(x_1) \ \phi(x_2) \ldots \phi(x_n) \rangle$ can be found by functional differentiation

$$\left\{ \frac{\delta^n G}{\delta \xi(x_1) \ \delta \xi(x_2) \ldots \delta \xi(x_n)} \right\}_{\xi=0} = i^n \ \langle \phi(x_1) \ \phi(x_2) \ldots \phi(x_n) \rangle \ . \quad (1.2)$$

Conversely, if all correlation functions are known the characteristic functional can be found from summation of its Taylor expansion

$$G[\xi(x)] = 1 + \sum_{n=1}^{\infty} \frac{i^n}{n!} \int_\Omega dx_1 \int_\Omega dx_2 \ldots \int_\Omega dx_n \ \langle \phi(x_1) \ \phi(x_2) \ldots \phi(x_n) \rangle$$

$$\times \ \xi(x_1) \ \xi(x_2) \ldots \xi(x_n) \ . \quad (1.3)$$

The random functions $\phi(x)$ will be called Gaussian random functions (with zero mean) if their correlation functions have the decomposition property

$$\langle \phi(x_1) \ \phi(x_2) \ldots \phi(x_{2l+1}) \rangle = 0 \ , \qquad (l = 0, 1, 2, \ldots) \ , \quad (1.4a)$$

$$\langle \phi(x_1) \ \phi(x_2) \ldots \phi(x_{2l}) \rangle = \sum \prod_{\alpha=1}^{l} \langle \phi(x_{i_\alpha}) \ \phi(x_{j_\alpha}) \rangle \ , \qquad (l = 1, 2, 3, \ldots) \ , \quad (1.4b)$$

where the sum runs over all the different way in which the $2l$ indices 1, 2,$\ldots$,$2l$ can be subdivided into $l$ unordered pairs $(i_1, j_1)$, $(i_2, j_2)$,$\ldots$, $(i_n, j_n)$. Hence, for Gaussian random functions all odd terms in (1.3) vanish, whereas the term with $n = 2l$ gives rise to

$$D_{2l} = \frac{(2l)!}{2^l l!} \quad (1.5)$$

identical contributions, each of magnitude

$$\frac{(-1)^l}{(2l)!} \left\{ \int_\Omega dx \int_\Omega dx' \ \xi(x) \ \langle \phi(x) \ \phi(x') \rangle \ \xi(x') \right\}^l \ . \quad (1.6)$$

Thus the characteristic functional of a Gaussian random function is given by

$$G[\xi(x)] = \exp \left\{ -\frac{1}{2} \int dx \int dx' \ \xi(x) \ \langle \phi(x) \ \phi(x') \rangle \ \xi(x') \right\} \ . \quad (1.7)$$

Note that this functional is uniquely determined by the two-point correlation function (the covariance).

Now consider the eigenvalue problem associated with the covariance

$$\int_{\Omega} \langle \phi(x) \, \phi(x') \rangle \, u_k(x') \, dx' = \lambda_k \, u_k(x) \ . \tag{1.8}$$

The eigenfunctions $u_k$ are assumed to form an orthonormal and complete set and the $\lambda_k$ are assumed to be real. The spectral representation of the functions $\phi(x)$ and $\xi(x)$ is defined by the expansions

$$\phi(x) = \sum_{\lambda_k > 0} a_k \, u_k(x) + i \sum_{\lambda_k < 0} a_k \, u_k(x) \ , \tag{1.9a}$$

$$\xi(x) = \sum_{\lambda_k > 0} b_k \, u_k(x) + i \sum_{\lambda_k < 0} b_k \, u_k(x) \ , \tag{1.9b}$$

where the $a_k$ and $b_k$ are real. Here the first (second) term on the right means a sum over those $\lambda_k$ which are positive (negative). Hence the real parts of $\phi$ and $\xi$ are composed of eigenfunctions with positive eigenvalues, and the imaginary parts are composed of eigenfunctions with negative eigenvalues. Substituting (1.9b) into (1.7) one finds

$$G\{b_k\} = \exp\left( -\frac{1}{2} \sum_{\lambda_k > 0} \lambda_k b_k^2 + \frac{1}{2} \sum_{\lambda_k < 0} \lambda_k b_k^2 \right) \ , \tag{1.10}$$

an expression which is finite when integrated over all $b_k$ from $-\infty$ to $+\infty$.

In the spectral representation, i.e., upon substitution of (1.9), Eq. (1.1) reads

$$G[\{b_k\}] = \langle \exp(i \sum_{\lambda_k > 0} a_k b_k - i \sum_{\lambda_k < 0} a_k b_k) \rangle \tag{1.11a}$$

$$= \left( \prod_k \int_{-\infty}^{+\infty} da_k \right) P[\{a_k\}] \exp(i \sum_{\lambda_k > 0} a_k b_k - i \sum_{\lambda_k < 0} a_k b_k) \ , \tag{1.11b}$$

where we denoted the probability density of the expansion coefficients $a_k$ by $P[\{a_k\}]$. Substitution of (1.10) shows that

$$P[\{a_k\}] = \prod_k (2\pi \mid \lambda_k \mid)^{-1/2} \exp\left( -\frac{a_k^2}{2 \mid \lambda_k \mid} \right) \ . \tag{1.12}$$

Note again that the integral of the probability density is finite (and equal to unity) when all $a_k$ are integrated from $-\infty$ to $+\infty$.

The continuous form of the probability density in function space can be obtained by substituting into the last equation the inverse of (1.9a)

$$a_k = \int \phi(x) \, u_k(x) \, dx \qquad \text{if} \qquad \lambda_k > 0 \ ,$$

$$a_k = \frac{1}{i} \int \phi(x) \, u_k(x) \, dx \qquad \text{if} \qquad \lambda_k < 0 \ . \tag{1.13}$$

This gives

$$P[\phi(x)] = N^{-1} \exp\left\{ -\frac{1}{2} \int_\Omega dx \int_\Omega dx' \, \phi(x) \, \langle \phi(x)\phi(x')\rangle^{-1} \, \phi(x') \right\} \ , \tag{1.14}$$

where the inverse of the covariance is denoted by

$$\langle \phi(x) \, \phi(x') \rangle^{-1} \equiv \sum_k \lambda_k^{-1} \, u_k(x) \, u_k(x') \ , \tag{1.15}$$

and where $N$ denotes the normalization integral

$$N = \int \exp\left\{ -\frac{1}{2} \int_\Omega dx \int_\Omega dx' \, \phi(x) \, \langle \phi(x)\phi(x') \rangle^{-1} \, \phi(x') \right\} d[\phi(x)] \ . \tag{1.16}$$

Of course, the same remarks apply here which were made in Sections 1.2 and 1.3 with respect to the Wiener integral. It should be especially kept in mind that an expression like

$$\langle F[\phi(x)] \rangle = N^{-1} \int F[\phi(x)] \tag{1.17}$$

$$\times \exp\left\{ -\frac{1}{2} \int_\Omega dx \int_\Omega dx' \, \phi(x) \, \langle \phi(x) \, \phi(x')\rangle^{-1} \, \phi(x') \right\} d[\phi(x)]$$

is either a formal notation for the spectral representation

$$\langle F[\{a_k\}] \rangle = \prod_k \left( \int_{-\infty}^{+\infty} da_k \right) F[\{a_k\}] \, P[\{a_k\}] \ , \tag{1.18}$$

or it is meant to indicate the limit $M \to \infty$ of the ratio of two multiple integrals which are obtained from (1.16, 17) by dividing $\Omega$ in a large number ($M$) of cells, and making $\phi(x)$ constant inside each cell. The treatment in this section followed Refs. I–37 and II–40.

## 6.2 Parametrization of the classical partition function

Consider a system of $N$ identical classical particles in a volume $\Omega$, interacting pairwise through a two-body potential. Denoting the positions of particles $i$ and $j$ by $\mathbf{r}_i$ and $\mathbf{r}_j$, their interaction energy $V(\mathbf{r}_i - \mathbf{r}_j)$ can often be written in a natural way as the sum of two terms.

$$V(\mathbf{r}_i - \mathbf{r}_j) = V_0(\mathbf{r}_i - \mathbf{r}_j) + V_1(\mathbf{r}_i - \mathbf{r}_j) \ . \qquad (2.1)$$

For example, $V_0$ could denote the potential of a hard core in the pair potential

$$V_0(\mathbf{r}_i - \mathbf{r}_j) = +\infty \qquad \text{if} \quad |\, \mathbf{r}_i - \mathbf{r}_j \,| < \sigma \ ,$$

$$= 0 \qquad \text{if} \quad |\, \mathbf{r}_i - \mathbf{r}_j \,| > \sigma \ . \qquad (2.2)$$

and $V_1$ could denote an attractive "tail" in the interaction. If this tail is weak and long-range (i.e., if the depth of $V_1$ is small compared to $k_B T$ and the range large compared to $\sigma$) the system is usually called a "van der Waals gas." We shall call the system in which the particles interact only through the potential $V_0$ the "reference system."

It is often useful to parametrize the term $V_1$ in the interaction by means of Gaussian random functions with zero mean and with a covariance

$$\langle \phi(\mathbf{r}) \, \phi(\mathbf{r}') \rangle = -\beta V_1(\mathbf{r} - \mathbf{r}') \ . \qquad (2.3)$$

This use of Gaussian random functions in classical statistical physics has been especially propagated by Siegert and his coworkers. A review of much of their work on the van der Waals gas can be found in Refs. 1,2. Applying the result (1.7) for the characteristic functional (1.1) to a function

$$\xi(\mathbf{r}) = -i \sum_{j=1}^{N} \delta(\mathbf{r} - \mathbf{r}_j) \ ,$$

where the $\mathbf{r}_1, \ldots, \mathbf{r}_N$ denote the coordinates of $N$ identical classical particles, one finds

$$\left\langle \exp\left\{ \sum_{j=1}^{N} \phi(\mathbf{r}_j) \right\} \right\rangle = \exp\left\{ -\frac{1}{2}\beta \sum_{i,j=1}^{N} V_1(\mathbf{r}_i - \mathbf{r}_j) \right\} \ . \qquad (2.4)$$

This parametrization of the Boltzmann factor was discovered by Kac [3].

The classical canonical partition function of these $N$ particles is given by the multiple integral

$$Z(N, \beta, \Omega) = \frac{1}{N!} \left( \frac{2\pi m}{\beta h^2} \right)^{\frac{1}{2}Nd} \int_\Omega d\mathbf{r}_1 \int_\Omega d\mathbf{r}_2 \ldots \int_\Omega d\mathbf{r}_N \qquad (2.5)$$

$$\times \exp\left\{ -\beta \sum_{i<j} V_0(\mathbf{r}_i - \mathbf{r}_j) - \beta \sum_{i<j} V_1(\mathbf{r}_i - \mathbf{r}_j) \right\} ,$$

where $m$ is the mass of the particles, $\beta = (k_B T)^{-1}$, $h$ is Planck's constant, $d$ the dimensionality of space and $\Omega$ the volume of the system. Combination of the last two equations gives, for the canonical partition function, the functional integral

$$Z(N, \beta, \Omega) = \exp\left\{\frac{1}{2}\beta\, V_1(0)N\right\} \langle Z_0(N, \beta, \Omega, -k_B T\phi)\rangle , \qquad (2.6)$$

where $Z_0(N, \beta, \Omega, \Phi)$ denotes the partition function of the reference system in an external potential $\Phi(\mathbf{r})$ which can be position dependent. What this means is that the long range part $V_1$ of the interaction between the particles can be replaced by an interaction of each single particle with an external field, followed by an averaging process over the external field. Note that only the finite part of the interaction in the van der Waals gas is parametrized by Gaussian random functions, whereas the hard core is treated as the reference system. This division of the interaction is forced upon us as it is impossible to parametrize a hard core potential by Gaussian random functions. Yet, in a case in which $V$ is everywhere finite, the division (2.1) of the potential can be done in any way one chooses, and will in practice be determined by calculational convenience.

For the grand canonical partition function

$$Z(z, \beta, \Omega) \equiv \sum_{N=1}^{\infty} Z(N, \beta, \Omega)\, z^N \qquad (2.7)$$

the corresponding functional integral is

$$Z(z, \beta, \Omega) = <Z_0(\zeta, \beta, \Omega, -k_B T\phi)> , \qquad (2.8)$$

where

$$\zeta = z \exp\left\{\frac{1}{2}\beta V_1(0)\right\} \qquad (2.9)$$

is a "renormalized" activity. In order to calculate the grand canonical partition function we specialize to the van der Waals gas and use the fact that $V_1$ represents a weak, long range interaction. Several alternative forms of the resulting simplication of the functional integral (2.8) can be found in Refs. 4, 5, II–38, II–40. The essential idea is as follows. If $V_1$ is long ranged with a scale $\gamma^{-1}$ one can determine a length $l$ such that

$$\sigma \ll l \ll \gamma^{-1} , \qquad (2.10)$$

provided $\gamma$ is small enough (hence certainly in the limit $\gamma \downarrow 0$). Dividing $\Omega$

into cubic cells of volume $l^d$ the Gaussian random functions $\phi$, which also vary in space on a scale of order $\gamma^{-1}$, are practically constant within each cell. At the same time, the number of particles in a cell of volume $l^d$ is large enough that the grand canonical partition function for cell $j$, in which the field equals $\phi_j$, can be approximated by its bulk limit

$$Z_0(\zeta, \beta, l^d, -k_BT\phi_j) \cong \exp\left\{\beta l^d p_0 \left(\zeta e^{\phi_j}, \beta\right)\right\} \qquad (2.11)$$

where $p_0(z, \beta)$ denotes the grand canonical pressure of the reference system at activity $z$ and temperature $T$. Hence, provided $\sigma\gamma \ll 1$, one has

$$Z_0(\zeta, \beta\ \Omega, -k_BT\phi) \cong \exp\left\{\beta \int_\Omega p_0(\zeta e^{\phi(\mathbf{r})}, \beta)\, d\mathbf{r}\right\} \qquad (2.12)$$

and the functional integral (2.8) becomes

$$Z(z, \beta, \Omega) \cong \left\langle \exp\left\{\beta \int_\Omega p_0(\zeta e^{\phi(\mathbf{r})}, \beta)\, d\mathbf{r}\right\}\right\rangle . \qquad (2.13)$$

In the remainder of this chapter we shall apply the last equation to an often studied model system.

## 6.3   The model of Kac, Uhlenbeck and Hemmer

Now specialize the theory of the previous section to a one-dimensional van der Waals gas with the special form

$$V_1(r - r') = -\frac{1}{2}\, w_0\, \gamma \exp\{-\gamma\, |\, r - r'|\} \qquad (3.1)$$

for the attractive tail. The covariance (2.3) of the Gaussian random functions is

$$\langle \phi(r)\ \phi(r')\rangle = \frac{1}{2}\beta w_0\, \gamma \exp\{-\gamma\, |r - r'|\} . \qquad (3.2)$$

It is straightforward to verify that

$$\left(\frac{d^2}{dr^2} - \gamma^2\right) \langle \phi(r)\ \phi(r')\ \rangle = -\beta w_0\gamma^2\ \delta(r - r') ; \qquad (3.3)$$

hence the inverse (1.15) of the covariance is given by a differential operator times a delta function

$$\langle \phi(r)\ \phi(r')\rangle^{-1} = (\beta w_0\gamma^2)^{-1} \left(\frac{d^2}{dr^2} - \gamma^2\right) \delta(r - r') . \qquad (3.4)$$

Substitution of the last expression and (1.16, 17) into (2.13) leads to a path-integral expression for the grand canonical partition function of this model

$$Z(z, \beta, L) = N^{-1} \int \exp\left\{ -(2\beta w_0 \gamma^2)^{-1} \int_{-L/2}^{+L/2} \left(\frac{d\phi}{dr}\right)^2 dr \right.$$
$$\left. - \int_{-L/2}^{+L/2} A(\phi)\, dr \right\} d[\phi(r)] \ . \qquad (3.5)$$

Here $L$ denotes the length of the system and

$$A(\phi) = \frac{\phi^2}{2\beta w_0} - \beta\, p_0(\zeta e^\phi, \beta) \ . \qquad (3.6)$$

The normalization is given by a similar path integral with $A$ replaced by $A_0 = \phi^2/2\beta w_0$.

The path integrals just found are Wiener integrals of the type (I.4.5) discussed in Chapter I, apart from the fact that in this case we also integrate over the values $\phi(L/2)$ and $\phi(-L/2)$ of the random function at the boundary of the container, whereas these values are kept constant in the Wiener integral. If, for example, we impose periodic boundary conditions

$$\phi(L/2) = \phi(-L/2) \equiv \phi \qquad (3.7)$$

on the random functions, then $Z$ can be expressed in terms of the solutions of two differential equations of the type (I.4.8)

$$Z(z, \beta, L) = \frac{\displaystyle\int G_A\left(\phi, +\frac{L}{2} \mid \phi, -\frac{L}{2}\right) d\phi}{\displaystyle\int G_{A_0}\left(\phi, +\frac{L}{2} \mid \phi, -\frac{L}{2}\right) d\phi} \ . \qquad (3.8)$$

The propagator $G_A$ can be expressed in the orthonormal eigenfunctions $f_n(\phi)$ and eigenvalues $E_n$ of the corresponding eigenvalue problem

$$\left[ -\frac{1}{2}\beta w_0 \gamma^2 \frac{d^2}{d\phi^2} + A(\phi) \right] f_n(\phi) = E_n f_n(\phi) \qquad (3.9)$$

by means of the bilinear expansion

$$G_A\left(\phi', +\frac{L}{2} \mid \phi, -\frac{L}{2}\right) = \sum_n f_n(\phi') f_n^*(\phi) \exp(-E_n L) \ . \qquad (3.10)$$

In the same way the propagator $G_{A_0}$ in the denominator of (3.8) can be written as

$$G_{A_0}\left(\phi', +\frac{L}{2} \mid \phi, -\frac{L}{2}\right) = \sum_n f_n^{(0)}(\phi') f_n^{(0)*}(\phi) \exp(-E_n^{(0)}L) , \qquad (3.11)$$

$$\left[-\frac{1}{2}\beta w_0 \gamma^2 \frac{d^2}{d\phi^2} + \frac{\phi^2}{2\beta w_0}\right] f_n^{(0)}(\phi) = E_n^{(0)} f_n^{(0)}(\phi) . \qquad (3.12)$$

This eigenvalue problem is essentially equivalent to the one for the harmonic oscillator in quantum mechanics, and one has

$$E_n^{(0)} = \left(n + \frac{1}{2}\right)\gamma , \qquad n = 0, 1, 2, \ldots \qquad (3.13)$$

Combining the last six equations, and taking the thermodynamic limit $L \to \infty$, one finds that the grand canonical pressure of the model of Kac, Uhlenbeck and Hemmer is entirely determined by the lowest eigenvalue of (3.9)

$$\beta p(z, \beta) = \lim_{L \to \infty} \frac{1}{L} \ln Z(z, \beta, L) = \frac{1}{2}\gamma - E_0 . \qquad (3.14)$$

In the derivation of this rigorous result we have followed Ref. I–37, II–38 and II–40. Helfand [6, 7] has analysed the phase transition in a one-dimensional Ising model with long-range interactions using analogous methods. Van Kampen [5] gives a similar analysis for the free energy. Of course, the serious student of this model system should also read the original papers of Kac, Uhlenbeck and Hemmer [8, 9, 10].

Let us now qualitatively discuss the physical interpretation of this result, following the nine references just quoted. Equation (3.6) shows that we need the explicit form of the grand canonical pressure of a one-dimensional gas consisting of hard rods of length $\sigma$. This is called the Tonks gas. Its thermodynamic functions are fully know and most of the relevant papers have been reprinted in the monograph by Mattis and Lieb [11]. For the sake of completeness we derive some of the main results.

Consider $N$ hard rods, each of length $\sigma$, on an interval of length $L$. If their midpoints have coordinates $x_1, x_2, \ldots, x_N$ the canonical partition function can be written in the form of a convolution integral

$$Z_0(N, \beta L) = \left(\frac{2\pi m}{\beta h^2}\right)^{N/2} \int_0^L dx_1 \int_{x_1+\sigma}^L dx_2 \ldots \int_{x_{N-1}+\sigma}^L dx_N . \qquad (3.15)$$

Its Laplace transform

$$Z_0(N, \beta, s) \equiv \int_0^\infty Z_0(N, \beta, L) \exp(-sL) \, dL \qquad (3.16)$$

is a product of two factors $s^{-1}$ and $(N-1)$ factors $s^{-1} e^{-s\sigma}$:

$$Z_0(0, \beta, s) = s^{-1} , \qquad (3.17a)$$

$$Z_0(1, \beta, s) = \lambda_B^{-1} s^{-2} , \qquad (3.17b)$$

$$Z_0(N, \beta, s) = \lambda_B^{-N} s^{-N-1} e^{-(N-1)\sigma s}, \qquad (N \geq 2) , \qquad (3.17c)$$

where

$$\lambda_B \equiv \left( \frac{h}{2\pi m k_B T} \right)^{1/2} \qquad (3.18)$$

is the thermal wavelength of the particles. Hence the generating function is found to be given by

$$Z_0(z, \beta, s) \equiv \sum_{N=0}^\infty Z_0(N, \beta, s) z^N$$

$$= \frac{1}{s} \left( 1 + \frac{z}{\lambda_B s - z \, e^{-\sigma s}} \right) . \qquad (3.19)$$

This of course enables us to calculate the grand canonical partition function by taking the inverse Laplace transform

$$Z_0(z, \beta, L) = \frac{1}{2\pi i} \int_C \frac{1}{s} \left( 1 + \frac{z}{\lambda_B s - z \, e^{-\sigma s}} \right) e^{+sL} \, ds . \qquad (3.20)$$

The contour of integration $C$ is a straight line, parallel to the imaginary $s$ axis, to the right of $s = 0$ and all the solutions $s^*$ of

$$z = \lambda_B s^* \, e^{\sigma s^*} . \qquad (3.21)$$

It is easy to verify that for real positive values of $z$ this equation has only one solution $s^*(z)$ which is situated on the positive real $s$ axis. The contour can be closed by a large semi-circle in the left half of the complex $s$-plane. Applying Cauchy's theorem and taking the residue at $s^*$ we find for the leading term in the grand canonical partition function

$$Z_0(z, \beta, L) \cong \frac{z/s^*}{\lambda_B + \lambda_B \sigma s^*} \exp\{s^*(z) L\} , \qquad (L \to \infty) . \qquad (3.22)$$

Hence the grand canonical pressure of the Tonks gas is given by

$$\beta p_0(z, \beta) = \lim_{L \to \infty} \frac{1}{L} \ln Z_0 = s^*(z) \tag{3.23}$$

The number density follows from

$$\rho_0(z, \beta) = z \frac{\partial}{\partial z} \beta p_0(z, \beta) = z \frac{\partial s^*}{\partial z}, \tag{3.24a}$$

or using (3.21)

$$\rho_0(z, \beta) = \frac{s^*(z)}{1 + \sigma s^*(z)}. \tag{3.24b}$$

Solving this equation with respect to $s^*$ and substituting into (3.23) we obtain the equation of state

$$\beta p_0(\rho) = \frac{\rho}{1 - \sigma \rho}. \tag{3.25}$$

Finally, the relation between density and activity of this system is found from (3.21, 23, 25)

$$z = \frac{\lambda_B \rho}{1 - \sigma \rho} \exp\left(\frac{\sigma \rho}{1 - \sigma \rho}\right). \tag{3.26}$$

After this intermezzo about the properties of the Tonks gas we return to the physical interpretation of expression (3.14) for the grand canonical pressure of the model of Kac, Uhlenbeck and Hemmer. It turns out that the physical properties of this model are related to the minima of the function $A(\phi)$ defined by (3.6). These minima are among the solutions of

$$\left(\frac{\partial A}{\partial \phi}\right)_{\beta, \zeta} = \frac{\phi}{\beta w_0} - \rho_0(\zeta e^\phi, \beta) = 0. \tag{3.27}$$

In order to locate the minima we note that (3.26) can be written in the form

$$\ln z = \frac{\sigma \rho}{1 - \sigma \rho} + \ln\left(\frac{\lambda_B \rho}{1 - \sigma \rho}\right). \tag{3.28}$$

Hence the curve which represents $\ln z$ as a function of $\rho$ is monotonically increasing on the physically interesting interval $0 \le \rho < \sigma^{-1}$, and the curve which gives $\rho_0(z, \beta)$ as a function of $\ln z$ is monotonically increasing for $-\infty < \ln z < +\infty$. A straightforward but tedious calculation shows that there is a unique point of inflection at $\ln z = \frac{1}{2} - \ln 2$, where $\rho_0 = (3\sigma)^{-1}$. The slope at this point equals

$$\max \frac{\partial \sigma_0}{\partial \ln z} = \frac{4}{27 \sigma} . \tag{3.29}$$

Returning to (3.27) this implies that the maximum slope of the term $\rho_0(\zeta e^\phi, \beta)$ as a function of $\phi$ equals $4/27\sigma$. As the term $\phi/\beta w_0$ can be represented by a straight line with slope $k_B T/w_0$ we have found that only two possibilities occur:
(a)  For $k_B T/w_0 > 4/27\sigma$ the function $A(\phi)$ has only one minimum $\phi_0$.
(b)  For $k_B T/w_0 < 4/27\sigma$ there always exists a range of values of the fugacity $z$ such that $A(\phi)$ has two minima, at $\phi_g$ and $\phi_l$ say. The transition between these two regions occurs at a temperature

$$T_w = \frac{4w_0}{27\sigma k_B} . \tag{3.30}$$

At fugacity $z = z_0(T)$ the two wells have the same depth: $A(\phi_g(z_0), T) = A(\phi_l(z_0), T)$. In both cases (a, b) the physical properties of the system depend on whether or not one takes the limit $\gamma \to 0$.
(1)  $\gamma \to 0$: The eigenvalue problem (3.9) has the form of the eigenvalue problem for a quantum mechanical particle with coordinate $\phi$ moving in an external potential $A(\phi)$. In the limit $\gamma \to 0$, which corresponds to the classical limit of the quantum mechanical particle, classical mechanics is rigorous and the ground state $E_0$ is the absolute minimum of $A(\phi)$. Equation (3.14) now gives

$$\beta p(z, \beta) = -A(\phi_0) , \qquad\qquad (T > T_W) , \quad (3.31a)$$

$$\beta p(z, \beta) = \max[-A(\phi_g) , -A(\phi_l)] , \qquad (T < T_W) . \quad (3.31b)$$

Combining this result with the explicit form (3.6) of $A(\phi)$ and with the definition (3.27) of $\phi_0$, $\phi_g$, $\phi_l$ as the local minima of $A(\phi)$ it is straightforward to verify that (3.31) is the grand canonical form of the van der Waals equation of state

$$p(\rho, T) = p_0(\rho, T) - \frac{1}{2} w_0 \rho^2 , \tag{3.32}$$

including the Maxwell construction (here $\rho = N/L$ denotes the number density). Thus the system has a gas phase and a liquid phase for temperatures below $T_W$, and one phase only for temperatures above $T_W$. Clearly $T_W$ is the critical temperature.
(2)  $\gamma > 0$: The eigenvalue problem (3.9) now presents us with a genuine

quantum mechanical problem, which can only be solved using some approximation technique. Using the WKB method of semiclassical quantum mechanics one finds for $T < T_W$ the expression

$$\beta p(z, \beta) = \frac{1}{2}\gamma - \frac{1}{2}(E_g + E_l) + \frac{1}{2}\sqrt{(E_g - E_l)^2 + \varepsilon^2} \, , \qquad (3.33a)$$

$$\varepsilon^2 = \frac{4}{\pi^2}(E_g - A(\phi_g))(E_l - A(\phi_l))\exp(-2c/\gamma) \, . \qquad (3.33b)$$

Here $c$ denotes a positive, $\gamma$-independent number and $E_g$ and $E_l$ denote the ground states of the one-well eigenvalue problems, i.e., $E_g$ is the ground state of the potential well in $A(\phi)$ around $\phi_g$ neglecting tunneling through the barrier between $\phi_g$ and $\phi_l$, and $E_l$ is defined similarly. Note that use of the WKB approximation implies that (3.33) is the leading term of an asymptotic expansion valid for small $\gamma$.

In this case of small, but positive $\gamma$, all the quantities $E_g$, $E_l$, $\phi_g$, $\phi_l$ and $c$ are analytic functions of the temperature $T$ and the fugacity $z$. Thus the grand canonical pressure is an analytic function of $z$ and the system will show no phase transition provided $\gamma > 0$. However, if $\gamma \to 0$ one has $E_g \to A(\phi_g)$, $E_l \to A(\phi_l)$; hence $\varepsilon \to 0$ and the square root in (3.33a) approaches the non-analytic function $|A(\phi_g) - A(\phi_l)|$, thereby transforming (3.33) into (3.31b). Note that the absence of a phase transition for $\gamma > 0$ is due to the presence of the $\varepsilon^2$ term in (3.33a). This term corresponds to the tunneling of the fictitious quantum mechanical particle through the potential barrier, as was shown by our discussion of the hopping paths approximation in Section 2.4.

**References**

[1]  A.J.F. Siegert, in *Statistical Mechanics at the Turn of the Decade*, E.G.D. Cohen, eds. (Marcel Dekker, New York, 1971) p. 145–174.

[2]  J.B. Jalickee, A.J.F. Siegert and D.J. Vezzetti, *J. Math. Phys.* **10** (1969) 1442.

[3]  M. Kac, *Phys. Fluids* **2** (1959) 8.

[4]  N.G. van Kampen, *Phys. Rev.* **135** (1964) A362.

[5]  N.G. van Kampen, *Physica* **48** (1970) 313.

[6]  E. Helfand, *J. Math. Phys.* **5** (1964) 127.

[7]  E. Helfand, in *The Equilibrium Theory of Classical Fluids*, H.L. Frisch and J.L. Lebowitz, eds. (Benjamin, New York, 1964) p. III 41.

[8]  M. Kac, G.E. Uhlenbeck and P.C. Hemmer, *J. Math. Phys.* **4** (1963) 216.

[9]  G.E. Uhlenbeck, P.C. Hemmer and M. Kac, *J. Math. Phys.* **4** (1963) 229.

[10] P.C. Hemmer, M. Kac and G.E. Uhlenbeck, *J. Math. Phys.* **5** (1964) 60.

[11] E.H. Lieb and D.C. Mattis, *Mathematical Physics in One Dimension*, (Academic Press, New York, 1966).

# VII.   QUANTUM STATISTICAL PHYSICS

In this chapter we discuss the equilibrium statistical mechanics of quantum systems, i.e., systems consisting of interacting bosons or interacting fermions. First we show how the partition function for such a system can be represented in several alternative ways by a path integral or by an integral over complex Gaussian random functions. These representations have the advantage that the quantum mechanical partition function is entirely expressed in terms of "classical" concepts such as c-number fields and integrals over c-number fields. We shall derive these representations in such a way that no operators, but only c-numbers appear in all the intermediate steps. In this way an entirely operator free form of the quantum statistics of interacting particles is developed. In the present chapter only rigorous results are derived. The application of the various approximation schemes to these path integrals leads to interesting physics: this will be the subject of the next two chapters. In all developments in this chapter the quantal character of the particles, as well as the (anti) symmetry of their wave functions will be taken into account rigorously; however, the degree of freedom corresponding to their spin will be ignored.

## 7.1   The Feynman path integral for the partition function

The canonical partition function of $N$ quantal particles (without any symmetry conditions on their wavefunctions!) in a volume $\Omega$ is usually represented by the trace of the density operator

$$Z(N, \beta, \Omega) = \text{Tr} \exp(-\beta H) \ . \tag{1.1}$$

Here $H$ is the Hamiltonian, which we shall assume to consist of the sum of a kinetic energy term,

$$-\frac{\hbar^2}{2m} \sum_{j=1}^{N} \Delta_j \equiv -\frac{\hbar^2}{2m} \sum_{j=1}^{N} \left( \frac{\partial^2}{\partial x_j^2} + \frac{\partial^2}{\partial y_j^2} + \frac{\partial^2}{\partial z_j^2} \right) \ ,$$

an interaction of each particle with an external field

$$\sum_{j=1}^{N} W(\mathbf{r}_j) \ ,$$

and a sum of two-particle interactions

$$\sum_{1 \leq i < j \leq N} V(\mathbf{r}_i - \mathbf{r}_j) \ .$$

Hence one has for the Hamiltonian operator

$$H = -\frac{\hbar^2}{2m} \sum_{j} \Delta_j + \sum_{j} W(\mathbf{r}_j) + \sum_{i<j} V(\mathbf{r}_i - \mathbf{r}_j) \ . \tag{1.2}$$

The trace is extended over a complete orthornormal set of functions in the $N$-particle Hilbert space.

If the particles are bosons Eq. (1.1) has to be replaced by

$$Z_{\text{B}}(N, \beta, \Omega) = \frac{1}{N!} \sum_{P} \text{Tr} \exp(-\beta H) P \ , \tag{1.3a}$$

and if they are fermions by

$$Z_{\text{F}}(N, \beta, \Omega) = \frac{1}{N!} \sum_{P} (\text{sgn } P) \text{Tr} \exp(-\beta H) P \ . \tag{1.3b}$$

In the last two equations $P$ denotes a permutation operator in the Hilbert space of $N$ particles, which can be defined by

$$P\psi(\mathbf{r}_1, \mathbf{r}_2, \ldots, \mathbf{r}_N) = \psi(\mathbf{r}_{P1}, \mathbf{r}_{P2}, \ldots, \mathbf{r}_{PN}) \ . \tag{1.4}$$

Here $\psi$ is any $N$-particle wave function and $(P1, P2, \ldots, PN)$ is a permutation of the indices $(1, 2, \ldots, N)$. The sign of $P$, denoted by sgn $P$ in (1.3b) and by $(-1)^P$ by some other authors, is $+1$ if $P$ is an even permutation and $-1$ if $P$ is an odd permutation.

As is well known, the trace of a matrix is an invariant. Suppose we evaluate the trace in the representation in which the position operators $\mathbf{r}_1$, $\mathbf{r}_2, \ldots, \mathbf{r}_N$ of the particles are simultaneously diagonal. Denoting these eigenfunctions by $|\mathbf{r}_1, \mathbf{r}_1, \ldots, \mathbf{r}_N\rangle$ Eq. (1.3a) becomes

$$Z_B(N, \beta, \Omega) = \frac{1}{N!} \sum_P \int d^{3N}r \, \langle \mathbf{r}_{P1}, \mathbf{r}_{P2}, \ldots, \mathbf{r}_{PN}|$$

$$\times \exp(-\beta H) \, | \, \mathbf{r}_1, \mathbf{r}_2, \ldots, \mathbf{r}_N \rangle \, , \qquad (1.5)$$

where the integral symbol indicates an integration of each of the variables $\mathbf{r}_1, \mathbf{r}_2, \ldots, \mathbf{r}_N$ over the volume $\Omega$. The matrix element is a special case of the $N$-particle propagator $G$ which is defined to be zero for $\beta'' < \beta'$, and which is given by

$$G(\mathbf{r}_1'', \mathbf{r}_2'', \ldots, \mathbf{r}_N'', \beta'' \, | \, \mathbf{r}_1', \mathbf{r}_2', \ldots, \mathbf{r}_N', \beta') \equiv$$

$$\langle \mathbf{r}_1'', \mathbf{r}_2'', \ldots, \mathbf{r}_N'' | \exp\{-(\beta'' - \beta')H\} | \mathbf{r}_1', \mathbf{r}_2', \ldots, \mathbf{r}_N' \rangle \qquad (1.6)$$

for $\beta'' > \beta'$. By direct substitution of (1.2) one easily verifies that this propagator satisfies the Bloch equation

$$\left[ \frac{\partial}{\partial \beta''} - \frac{\hbar^2}{2m} \sum_j \Delta_j'' + \sum_j W(\mathbf{r}_j'') + \sum_{i<j} V(\mathbf{r}_i'' - \mathbf{r}_j'') \right] \qquad (1.7)$$

$$\times \, G(\mathbf{r}_1'', \mathbf{r}_2'', \ldots, \mathbf{r}_N'', \beta'' \, | \, \mathbf{r}_1', \mathbf{r}_2', \ldots, \mathbf{r}_N', \beta') = \delta(\beta'' - \beta') \prod_j \delta(\mathbf{r}_j'' - \mathbf{r}_j') \, ,$$

where $\Delta_j''$ denotes the Laplace operator with respect to the coordinates $x_j''$, $y_j''$ and $z_j''$. As this is a generalization of the differential equation (I.4.8) the propagator can be represented by a Wiener intergral which is the generalization of (I.4.5)

$$G(\mathbf{r}_1'', \mathbf{r}_2'', \ldots, \mathbf{r}_N'', \beta'' \, | \, \mathbf{r}_1', \mathbf{r}_2', \ldots, \mathbf{r}_N', \beta') = \qquad (1.8)$$

$$\int_{\mathbf{r}_1', \beta'}^{\mathbf{r}_1'', \beta''} d[\mathbf{r}_1(\tau)] \int_{\mathbf{r}_2', \beta'}^{\mathbf{r}_2'', \beta''} d[\mathbf{r}_2(\tau)] \ldots \int_{\mathbf{r}_N', \beta'}^{\mathbf{r}_N'', \beta''} d[\mathbf{r}_N(\tau)]$$

$$\times \exp\left\{ -\frac{m}{\cdot 2\hbar^2} \int_{\beta'}^{\beta''} \sum_j \left[ \frac{d\mathbf{r}_j}{d\tau} \right]^2 d\tau - \int_{\beta'}^{\beta''} \sum_j W(\mathbf{r}_j)d\tau - \int_{\beta'}^{\beta''} \sum_{i<j} V(\mathbf{r}_i - \mathbf{r}_j)d\tau \right\} \, .$$

Substitution of this result into (1.5) one finds the rather formidable looking expression

$$Z_B(N, \beta, \Omega) = \frac{1}{N!} \sum_P \int d^{3N}r \int_{\mathbf{r}_1,0}^{\mathbf{r}_{P1},\beta} d[\mathbf{r}_1(\tau)] \int_{\mathbf{r}_2,0}^{\mathbf{r}_{P2},\beta} d[\mathbf{r}_2(\tau)]$$

$$\cdots \int_{\mathbf{r}_N,0}^{\mathbf{r}_{PN},\beta} d[\mathbf{r}_N(\tau)] \exp\left(-\int_0^\beta F \, d\tau\right) . \tag{1.9}$$

Here the spatial integrations are extended over the volume $\Omega$ and the functional $F$ is given by

$$F = \frac{m}{2\hbar^2} \sum_j \left[\frac{d\mathbf{r}_j}{d\tau}\right]^2 + \sum_j W(\mathbf{r}_j(\tau)) + \sum_{i<j} V(\mathbf{r}_i(\tau) - \mathbf{r}_j(\tau)) . \tag{1.10}$$

To summarize what (1.9) means we give the following recipe for a numerical calculation of the Bose partition function: (1) Place the particles at $N$ positions inside the volume $\Omega$. (2) Let them move, inside the volume, till at "time" $\beta$ their positions are a permutation $P$ of their initial positions. (3) Calculated the weight $\exp(-\int F \, d\tau)$. (4) Sum over all paths, integrate over all initial positions and sum over all permutations! For fermions one gets an extra factor sgn $P$ so in an even more abbreviated, but hopefully obvious, notation

$$Z_F(N, \beta, \Omega) = \frac{1}{N!} \sum_P (\text{sgn } P) \int d^{3N}r \int_{\mathbf{r},0}^{P\mathbf{r},\beta} d^N[\mathbf{r}(\tau)] \exp\left(-\int_0^\beta F \, d\tau\right) . \tag{1.11}$$

These results are due to Feynman. In subsequent papers [120–128] Feynman, ter Haar, Kikuchi and others used (1.9, 10) as the starting point for a theory of the $\lambda$ transition in $^4$He. In the next section we shall evaluate $Z_B$ and $Z_F$ explicitly for the case of non-interacting particles.

## 7.2  The ideal Bose gas and the ideal Fermi gas

In order to demonstrate how the path integrals (1.9) and (1.11) "work" we consider the case in which the particles do not interact with each other. In this case the trajectories of the particles are independent, apart from the fact that the endpoint of the trajectory of particle $i$ is the starting point of particle $Pi$. In order to take advantage of this fact it is convenient to decompose the permutations into cycles. Every permutation $P$ can be written uniquely as the product of $l$ cycles

$$P = C(s_1) \, C(s_2) \ldots C(s_l) , \tag{2.1}$$

where $C(s)$ denotes a cyclic permutation of $s \geq 1$ elements and where

$$\sum_{k=1}^{l} s_k = N \ . \tag{2.2}$$

It is clear that every cycle of $s$ elements will contribute a factor

$$f_s[W] = \int_\Omega d^3r_1 \int_\Omega d^2r_2 \ldots \int_\Omega d^3r_s$$

$$\times \int_{\mathbf{r}_1,0}^{\mathbf{r}_2,\beta} d[\mathbf{r}_1(\tau)] \int_{\mathbf{r}_2,0}^{\mathbf{r}_3,\beta} d[\mathbf{r}_2(\tau)] \ldots \int_{\mathbf{r}_s,0}^{\mathbf{r}_1,\beta} d[\mathbf{r}_s(\tau)]$$

$$\times \exp\left\{-\frac{m}{2\hbar^2} \int_0^\beta \sum_{j=1}^{s} \left[\frac{d\mathbf{r}_j}{d\tau}\right]^2 d\tau - \int_0^\beta \sum_{j=1}^{s} W(\mathbf{r}_j(\tau))d\tau\right\} \tag{2.3}$$

to the term in (1.9, 11) which corresponds with the permutation $P$. Hence

$$f_s[W] = \int_\Omega d^3r_1 \int_\Omega d^3r_2 \ldots \int_\Omega d^3r_s \, G_W(\mathbf{r}_1, \beta \mid \mathbf{r}_s, 0) \, G_W(\mathbf{r}_s, \beta \mid \mathbf{r}_{s-1}, 0)$$

$$\ldots G_W(\mathbf{r}_2, \beta \mid \mathbf{r}_1, 0) \ , \tag{2.4}$$

where

$$G_W(\mathbf{r}'', \tau'' \mid \mathbf{r}', \tau') \equiv \int_{\mathbf{r}',\tau'}^{\mathbf{r}'',\tau''} \exp\left\{-\frac{m}{2\hbar^2} \int_{\tau'}^{\tau''} \left(\frac{d\mathbf{r}}{d\tau}\right)^2 d\tau - \int_{\tau'}^{\tau''} W(\mathbf{r}(\tau)) \, d\tau\right\}$$

$$\times d[\mathbf{r}(\tau)] \ . \tag{2.5}$$

Using once more the results of Chapter I we see that $G_W$ can be evaluated explicitly in the form

$$G_W(\mathbf{r}'', \tau'' \mid \mathbf{r}', \tau') = \sum_n U_n(\mathbf{r}'') \, U_n^*(\mathbf{r}') \exp\{-E_n(\tau'' - \tau')\} \ , \tag{2.6}$$

where

$$-\frac{\hbar^2}{2m}\triangle U_n + W(\mathbf{r}) \, U_n = E_n U_n \ . \tag{2.7}$$

Hence the $E_n$ are the eigenvalues and the $U_n$ the orthonormal eigenfunctions of one particle in the volume $\Omega$. Substituting (2.6) into (2.4) one finds the simple expression

$$f_s[W] = \sum_n \exp(-E_n s\beta) \tag{2.8}$$

which shows that $f_s$ only depends on the spectrum of the external potential $W$, but not on the form of the eigenfunctions.

Next one notes that the number of permutations of $N$ elements which consist of $n_s = 0, 1, 2, \ldots$ cycles of $s = 1, 2, 3, \ldots$ elements is given by

$$C(N, \{n_s\}) = \frac{N!}{\displaystyle\prod_{s=1}^{\infty} (n_s! s^{n_s})} , \qquad (2.9)$$

where of course the numbers $n_s$ are restricted by the condition

$$\sum_{s=1}^{\infty} s n_s = N . \qquad (2.10)$$

Substitution of (2.9) into (1.9) gives, for the non-interacting case,

$$Z_B^{(0)} (N, \beta, \Omega) = \sum_{\{n_s\}}{}' \prod_{s=1}^{\infty} \frac{1}{n_s!} \left( \frac{1}{s} f_s \right)^{n_s} , \qquad (2.11)$$

where the prime indicates the constraints (2.10). The grand canonical partition function

$$Z_B^{(0)} (z, \beta, \Omega) \equiv \sum_{N=0}^{\infty} Z_B^{(0)} (N, \beta, \Omega) z^N \qquad (2.12)$$

now becomes explicitly

$$Z_B^{(0)} (z, \beta, \Omega) = \exp\left\{ \sum_{s=1}^{\infty} \frac{z^s}{s} f_s[W] \right\} . \qquad (2.13)$$

In the same way one finds for non-interacting fermions

$$Z_F^{(0)} (z, \beta, \Omega) = \exp\left\{ \sum_{s=1}^{\infty} (-1)^{s+1} \frac{z^s}{s} f_s[W] \right\} . \qquad (2.14)$$

Of course, in the present case of non-interacting particles the last two expressions can be evaluated explicitly with the use of (2.8). One finds

$$\sum_{s=1}^{\infty} \frac{z^s}{s} f_s[W] = \sum_n \sum_{s=1}^{\infty} \frac{1}{s} \{z \exp(-\beta E_n)\}^s$$

$$= -\sum_n \ln(1 - z \, e^{-\beta E_n}) . \qquad (2.15)$$

The grand canonical pressure for the ideal Bose gas is now found to be

$$\beta p_{\mathrm{B}}^{(0)} \ (z, \beta) = \lim_{\Omega \to \infty} \frac{1}{\Omega} \sum_n \ln(1 - z \ e^{-\beta E_n})^{-1} \ , \qquad (2.16)$$

and the density is

$$\rho_{\mathrm{B}}^{(0)} \ (z, \beta) = z \ \frac{\partial \beta p_{\mathrm{B}}^{(0)}}{\partial z} = \lim_{\Omega \to \infty} \frac{1}{\Omega} \sum_n \left( \frac{1}{z} \ e^{\beta E_n} - 1 \right)^{-1} \ . \qquad (2.17)$$

In the same way one finds for the ideal Fermi gas

$$\beta p_{\mathrm{F}}^{(0)} \ (z, \beta) = \lim_{\Omega \to \infty} \frac{1}{\Omega} \sum_n \ln\left( 1 + z e^{-\beta E_n} \right) \ , \qquad (2.18)$$

$$\rho_{\mathrm{F}}^{(0)} \ (z, \beta) = \lim_{\Omega \to \infty} \frac{1}{\Omega} \sum_n \left( \frac{1}{z} \ e^{\beta E_n} + 1 \right)^{-1} . \qquad (2.19)$$

The last four equations are the standard results for the non-interacting quantum gases, and lead to all their well known physical properties. The interested reader might consult any textbook on statistical physics to pursue the physics from here on.

## 7.3 Siegert's functional integral

If the particles interact through a pair potential the evaluation of the path integrals (1.9) and (1.11) becomes much more difficult. In this case the trajectories of two particles $i$ and $j$ are coupled through the term $V(\mathbf{r}_i(\tau) - \mathbf{r}_j(\tau))$ in (1.10). Yet, even in this case one can reduce the Feynman path integral considerably by replacing the pair interaction by an independent interaction of each particle with a fluctuating external field, followed by an averaging process over the external field.

Following Siegert [1] we introduce a Gaussian random field $\phi(\mathbf{r}, \tau)$ with zero mean value, and with a two-point correlation function

$$\langle \phi(\mathbf{r}, \tau) \ \phi(\mathbf{r}', \tau') \rangle = -V(\mathbf{r} - \mathbf{r}') \ \delta(\tau - \tau') \ . \qquad (3.1)$$

Using the explicit form (VI.1.7) of the characteristic functional (VI.1.1) for the field

$$\xi(\mathbf{r}, \tau) = -i \sum_{j=1}^{N} \delta(\mathbf{r} - \mathbf{r}_j(\tau)) \ , \qquad (3.2)$$

one finds the identity

$$\left\langle \exp\left\{ \int_0^\beta \sum_j \phi(\mathbf{r}_j(\tau), \tau) \ d\tau \right\} \right\rangle = \exp\left\{ -\frac{1}{2} \beta V(\mathbf{0}) N - \int_0^\beta \sum_{i<j} V(\mathbf{r}_i(\tau) - \mathbf{r}_j(\tau)) d\tau \right\} \ ,$$

$$(3.3)$$

which holds for any set of $N$ trajectories $r_1(\tau), r_2(\tau), \ldots, r_N(\tau)$. This, of course, enables us to "parametrize" the Boltzmann factor which appears in the Feynman path integrals (1.9, 11). For bosons one finds

$$Z_B(N, \beta, \Omega) = \exp\left\{+\frac{1}{2}\beta V(0)N\right\} \left\langle \frac{1}{N!} \sum_P \int d^{3N}r \int_{r,0}^{Pr, \beta} d^N[r(\tau)] \right.$$

$$\left. \times \exp\left(-\int_0^\beta F_\phi \, d\tau\right) \right\rangle ,  \qquad (3.4a)$$

where

$$F_\phi \equiv \frac{m}{2\hbar^2} \sum_j \left[\frac{dr_j}{d\tau}\right]^2 + \sum_j W(r_j(\tau)) - \sum_j \phi(r_j(\tau), \tau) ; \qquad (3.4b)$$

for fermions there is an extra factor (sgn $P$) in the sum over permutations.

The expression between the averaging brackets may be interpreted as the partition function of an ideal Bose gas in an external field given by $W(r) - \phi(r, \tau)$ which still depends on the parameter $\tau$. But this problem was studied in the last section. Going to the grand canonical partition function

$$Z_B(z, \beta, \Omega) \equiv \sum_{n=0}^{\infty} Z_B(N, \beta, \Omega) z^N \qquad (3.5)$$

we find from (2.13, 14) the functional averages

$$Z_B(z, \beta\,\Omega) = \left\langle \exp\left\{\sum_{s=1}^{\infty} \frac{\zeta^s}{s} f_s[W - \phi]\right\}\right\rangle , \qquad (3.6)$$

$$Z_F(z, \beta\,\Omega) = \left\langle \exp\left\{\sum_{s=1}^{\infty} (-1)^{s+1} \frac{\zeta^s}{s} f_s[W - \phi]\right\} , \qquad (3.7)$$

where

$$\zeta = z \exp\{\tfrac{1}{2}\beta V(0)\} . \qquad (3.8)$$

These functional integrals were first derived by Siegert [1]. They have a slightly more restricted range of applicability that Feynman's path integrals as (1.9, 11) hold if the pair interaction $V$ has a hard core, whereas, (3.6, 7) hold only for pair interactions that have a Fourier transform, which excludes the hard core. Yet, for physical applications this is no restriction at all because a hard core in a pair potential is itself unphysical and should be replaced by a core of "finite hardness," in which the interaction energy is large as compared to the thermal energy

$$V(\mathbf{r}) = V_0 >> k_B T \quad \text{if} \quad |\mathbf{r}| < \sigma . \tag{3.9}$$

### 7.4 Wiegel's functional integral

In going from Feynman's path integral to Siegert's functional integral one of the difficulties in evaluating the partition function of the interacting Bose gas has been eliminated: the two particle interaction, which couples the trajectories of the particles, has been replaced by an interaction of each individual particle with a fluctuating "time" dependent external field $\phi(\mathbf{r}, \tau)$ [cf. Refs. 2,3]. However, the exponentials in (3.6, 7), which we shall denote by

$$F_B[\phi] \equiv \sum_{s=1}^{\infty} \frac{\zeta^s}{s} f_s[W - \phi] , \tag{4.1}$$

$$F_F[\phi] \equiv \sum_{s=1}^{\infty} (-1)^{s+1} \frac{\zeta^s}{s} f_s[W - \phi] , \tag{4.2}$$

are still complicated functionals of the external field, which makes an approximate evaluation of the partition function starting from Siegert's functional integral rather awkward. In this section we show that the exponentials $F_B$ and $F_F$ are much easier to handle when they are written in the form of a Taylor series in $\phi$, with coefficients which are functions of $\zeta$. We shall follow the algebraic proof of Ref. [4]; a diagrammatic proof can be found in [5, 6].

The first step in the derivation consists of an expansion of $F_B[\phi]$ in a functional Taylor series around $\phi = 0$. Denoting $\mathbf{r}, \tau$ by $x$ this series has the form

$$F_B[\phi] = F_B[0] + \sum_{n=1}^{\infty} \frac{1}{n!} \int dx_1 \int dx_2 \ldots \int dx_n$$

$$\times \left\{ \frac{\delta^n F_B}{\delta\phi(x_1) \, \delta\phi(x_2) \, \ldots \, \delta\phi(x_n)} \right\}_{\phi=0} \phi(x_1) \, \phi(x_2) \ldots \phi(x_n) . \tag{4.3}$$

Here the $x$-integration extends over the volume $\Omega$ in $\mathbf{r}$ space and over the interval $0 < \tau < \beta$ in $\tau$ space. The term $F_B[0]$ can be brought outside the average in (3.6) and gives rise to a factor

$$Z_B^{(0)} (\zeta, \beta, \Omega) = \exp\{F_B[0]\} , \qquad (4.4)$$

which is the grand canonical partition function of the ideal Bose gas, calculated in Section 7.2, but at activity $\zeta$ instead of $z$.

The functional derivative of $F_B$ with respect of $\phi(y)$ follows by combination of (4.1) with (2.4)

$$\frac{\delta F_B}{\delta \phi(y)} = \sum_{s=1}^{\infty} \zeta^s \int d^{3s}r \left\{ \frac{\delta G_{W-\phi} (\mathbf{r}_1, \beta \mid \mathbf{r}_s, 0)}{\delta \phi(y)} \right\}$$

$$\times G_{W-\phi} (\mathbf{r}_s, \beta \mid \mathbf{r}_{s-1}, 0) \ldots G_{W-\phi} (\mathbf{r}_2, \beta \mid \mathbf{r}_1, 0) . \qquad (4.5)$$

Now, the functional derivative which appears here can easily be calculated from the integral equation (similar to I.4.10) which connects $G_{W-\phi}$ with $G_W$ and which we shall write in the symbolic notation

$$G_{W-\phi} = G_W + G_W \phi G_{W-\phi} \qquad (4.6)$$

In this formula $G_{W-\phi}$, $G_W$ and $\phi$ are interpreted as infinite matrices, the rows and columns of which are labelled by the continuous variable $x$. Note that $\phi$ is a diagonal matrix. The solution of the last equation is of course found by iteration

$$G_{W-\phi} = G_W + G_W \phi G_W + G_W \phi G_W \phi G_W + \ldots \qquad (4.7)$$

Hence the functional derivative is

$$\frac{\delta G_{W-\phi} (z \mid x)}{\delta \phi(y)} = G_W(x \mid y) G_W(y \mid z) + G_W(x \mid y) (G_W \phi G_W) (y \mid z)$$

$$+ (G_W \phi G_W) (x \mid y) G_W(y \mid z) + \ldots = G_{W-\phi}(x \mid y) G_{W-\phi}(y \mid z) . \qquad (4.8)$$

Using this result (4.5) becomes

$$\frac{\delta F_B}{\delta \phi(y)} = \sum_{s=1}^{\infty} \zeta^s \int d^{3s}r \, G_{W-\phi}(\mathbf{r}_1, \beta \mid y) \, G_{W-\phi}(y \mid \mathbf{r}_s, 0)$$

$$\times G_{W-\phi}(\mathbf{r}_s, \beta \mid \mathbf{r}_{s-1}, 0) \ldots G_{W-\phi}(\mathbf{r}_2, \beta \mid \mathbf{r}_1, 0) . \qquad (4.9)$$

We shall write this in the more suggestive notation

$$\frac{\delta F_B[\phi]}{\delta \phi(y)} = g_{W-\phi}(y \mid y) , \qquad (4.10)$$

where the continuous matrix $g_{W-\phi}$ is defined by

$$g_{W-\phi}(z|x) \equiv G_{W-\phi}(z|x) + \sum_{s=1}^{\infty} \zeta^s \int d^{3s} r\, G_{W-\phi}(z|\mathbf{r}_s, 0) \qquad (4.11)$$

$$\times\, G_{W-\phi}(\mathbf{r}_s, \beta\,|\mathbf{r}_{s-1}, 0) \ldots G_{W-\phi}(\mathbf{r}_2, \beta\,|\,\mathbf{r}_1, 0)\, G_{W-\phi}(\mathbf{r}_1, \beta\,|\,x) \ .$$

Note that the first term $G_{W-\phi}(z\,|\,x)$ will vanish when $z = x = y$ if we define

$$G_{W-\phi}(z\,|\,x) = 0 \quad \text{for} \quad t_z = t_x \ ; \qquad (4.12)$$

this completes the definition (I.4,7) for the case $t = t_0$.

In order to calculate the second functional derivative of the exponential $F_B$ we first note that $g_\phi$ has the same elegant differential property

$$\frac{\delta g_{W-\phi}(z\,|\,x)}{\delta\phi(y)} = g_{W-\phi}(z\,|\,y)\, g_{W-\phi}\,(y\,|\,x) \ , \qquad (4.13)$$

as follows from (4.8) and (4.11). Using this in combination with (4.10) one finds

$$\frac{\delta^2 F_B[\phi]}{\delta\phi(x_1)\,\delta\phi(x_2)} = g_{W-\phi}(x_1\,|\,x_2)\, g_{W-\phi}(x_2\,|\,x_1) \ . \qquad (4.14)$$

Hence the $n$-th term in the Taylor expansion (4.3) is found to equal

$$\int dx_1 \int dx_2 \ldots \int dx_n \left\{ \frac{\delta^n F_B}{\delta\phi(x_1)\,\delta\phi(x_2)\ldots\delta\phi(x_n)} \right\}_{\phi=0}$$

$$\times\, \phi(x_1)\,\phi(x_2)\ldots\phi(x_n)$$

$$= (n-1)! \int dx_1 \int dx_2\, \ldots \int dx_n\, \phi(x_1)\, g_W(x_1\,|\,x_2)\, \phi(x_2)\, g_W(x_2\,|\,x_3)$$

$$\ldots \phi(x_n)\, g_W(x_n\,|\,x_1) \qquad (4.15a)$$

$$= (n-1)!\, \mathrm{Tr}(\phi g_W)^n \ , \qquad (4.15b)$$

where $\mathrm{Tr}\, A$ denotes the trace of the matrix $A$, i.e., the sum of the matrix elements in the main diagonal.

Using the last formula one can now explicity sum the Taylor series (4.3)

$$F_B[\phi] = F_B[0] + \sum_{n=1}^{\infty} \frac{1}{n}\, \mathrm{Tr}(\phi g_W)^n = F_B[0] - \mathrm{Tr}\,\ln(1 - \phi g_W) \ . \qquad (4.16)$$

Substituting into Siegert's representation (3.6) one finds the functional integral

$$Z_B(z, \beta, \Omega) = Z_B^{(0)} (\zeta, \beta, \Omega) \langle \exp\{-\text{Tr} \ln(1 - \phi g_W(\zeta))\} \rangle \quad (4.17)$$

$$= Z_B^{(0)} (\zeta, \beta, \Omega) \langle \det(1 - \phi g_W(\zeta))^{-1} \rangle , \quad (4.18)$$

where the relation

$$\exp(\text{Tr } A) = \det \exp(A) \quad (4.19)$$

was used to derive (4.18) from (4.17). For fermions one finds

$$Z_F(z, \beta, \Omega) = Z_F^{(0)} (\zeta, \beta, \Omega) \langle \exp\{+ \text{Tr} \ln(1 - \phi g_W(-\zeta))\} \rangle$$

$$= Z_F^{(0)} (\zeta, \beta, \Omega) \langle \det(1 - \phi g_W(-\zeta)) \rangle . \quad (4.20)$$

Note that the same "grand canonical propagator" $g_W$ appears in these four functional integrals, but with argument $+\zeta$ for bosons, $-\zeta$ for fermions. Substituting the expansion (2.6) and $\phi = 0$ into the definition (4.11) one finds the explicit form for $g_W(\zeta)$

$$g_W(\mathbf{r}, \tau \mid \mathbf{r}_0, \tau_0) = \sum_n U_n(\mathbf{r}) U_n^*(\mathbf{r}_0) \frac{\exp\{-(\tau - \tau_0)E_n\}}{1 - \zeta\exp(-\beta E_n)}$$

$$(0 < \tau_0 < \tau < \beta) , \quad (4.21a)$$

$$g_W(\mathbf{r}, \tau \mid \mathbf{r}_0, \tau_0) = \sum_n U_n(\mathbf{r}) U_n^*(\mathbf{r}_0) \frac{\exp\{-(\tau - \tau_0)E_n\}}{1 - \zeta\exp(-\beta E_n)} \zeta\exp(-\beta E_n)$$

$$(0 < \tau \leq \tau_0 < \beta) . \quad (4.21b)$$

This function does not itself depend on the interaction, and characterizes the quantum character of the Bose and Fermi gasses.

### 7.5  Gaussian random functions with an asymmetric correlation function

In the present section, which is of the nature of a short mathematical intermezzo, we discuss the problem of how to construct Gaussian random functions with a covariance which is not symmetric. This problem cannot be solved with the theory of Section 6.1 which always leads to a symmetric covariance. In the next section we shall use our results to transform Wiegel's functional integral into Bell's functional integral, which will be a convenient starting point for an approximate theory of the interacting Bose fluid (cf. Chapters VIII and IX). We shall especially follow the methods of Refs. 5 and 6.

Let the correlation function of the Gaussian random functions be denoted by $g(x|x')$. In order to simulate its non-symmetric character $[g(x|x') \neq g(x'|x)]$ we introduce a complex valued random function $\psi(x)$ and its adjoint $\bar{\psi}(x)$. For the time being we do not yet specify what is meant by "adjoint." The Gaussian character of the random function is defined by a generalized form of the decomposition theorem

$$\langle \psi(x)\,\psi(x') \rangle = 0 \ , \tag{5.1a}$$

$$\langle \bar{\psi}(x)\,\bar{\psi}(x') \rangle = 0 \ , \tag{5.1b}$$

$$\langle \psi(x)\,\bar{\psi}(x') \rangle = g(x \mid x') \ , \tag{5.1c}$$

and

$$\left\langle \prod_{i=1}^{m} \{\psi(x_i)\} \prod_{j=1}^{n} \{\bar{\psi}(x_j')\} \right\rangle = 0 \ , \qquad (m \neq n) \ , \tag{5.2a}$$

$$= \sum_{P} \prod_{i=1}^{m} \langle \psi(x_i)\,\bar{\psi}(x_{Pi}') \rangle \qquad (m = n) \ , \tag{5.2b}$$

where the sum extends over all permutations $P$ of the $m$ numbers $i = 1, 2, \ldots, m$.

The characteristic functional is now defined by

$$G[\xi(x)] \equiv \left\langle \exp\{i \int [\psi(x)\bar{\xi}(x) + \bar{\psi}(x)\,\xi(x)]\,dx\} \right\rangle \ . \tag{5.3}$$

Expanding the exponential in a Taylor series, and applying the decomposition theorem to each term, one finds explicitly

$$G[\xi(x)] = 1 + \sum_{n=1}^{\infty} \frac{i^n}{n!} \int dx_1 \int dx_2 \ldots \int dx_n$$

$$\times \left\langle \prod_{j=1}^{n} \{\psi(x_j)\,\bar{\xi}(x_j) + \bar{\psi}(x_j)\,\xi(x_j)\} \right\rangle$$

$$= 1 + \sum_{l=1}^{\infty} \frac{(-1)^l}{(2l)!} \frac{(2l)!}{(l!)^2} \int dx_1 \int dx_2 \ldots \int dx_{2l}$$

$$\times \langle \psi(x_1) \ldots \psi(x_l)\,\bar{\psi}(x_{l+1}) \ldots \bar{\psi}(x_{2l}) \rangle$$

$$\times \bar{\xi}(x_1) \ldots \bar{\xi}(x_l)\,\xi(x_{l+1}) \ldots \xi(x_{2l})$$

$$= 1 + \sum_{l=1}^{\infty} \frac{(-1)^l}{l!} \left[ \int dx \int dx'\,\bar{\xi}(x)\,\langle \psi(x)\,\bar{\psi}(x') \rangle\,\xi(x') \right]^l$$

$$= \exp\left\{- \int dx \int dx' \; \bar{\xi}(x) \, g(x \mid x') \, \xi(x')\right\} .$$  (5.4)

This simple result leads us to introduce the "spectral" representation of the random functions $\bar{\psi}(x)$ in the following way. First, one solves for the right and left eigenfunctions associated with the asymmetric kernel $g$. Denote them by $\chi_k$ and $\bar{\chi}_k$:

$$\int g(x \mid x') \, \chi_k(x') \, dx' = \lambda_k \, \chi_k(x) ,$$  (5.5a)

$$\int \bar{\chi}_k(x) \, g(x \mid x') \, dx = \lambda_k \, \bar{\chi}_k(x') .$$  (5.5b)

Second, one expands $\psi$ and $\xi$ in the right eigenfunctions, $\bar{\psi}$ and $\bar{\xi}$ in the left eigenfunctions

$$\psi(x) = \sum_k a_k \, \chi_k(x) ,$$  (5.6a)

$$\xi(x) = \sum_k b_k \, \chi_k(x) ,$$  (5.6b)

$$\bar{\psi}(x) = \sum_k a_k^* \, \bar{\chi}_k(x) ,$$  (5.6c)

$$\bar{\xi}(x) = \sum_k b_k^* \, \bar{\chi}_k(x) ,$$  (5.6d)

where the star denotes the complex conjugate number. Note that (5.5b) and (5.6c, d) define what is meant by the "adjoint" random function. Using the orthonormality relation

$$\int \bar{\chi}_k(x) \, \chi_l(x) \, dx = \delta_{k,l}$$  (5.7)

and substituting (5.6b, d) into (5.4) one finds the spectral form of the characteristic functional

$$G(\{b_k\}) = \exp\left\{- \sum_k \lambda_k \, b_k^* \, b_k\right\} .$$  (5.8)

Introducing the real and imaginary parts of the complex expansion coefficients

$$a_k = a_k' + i \, a_k'' ,$$  (5.9a)

$$b_k = b_k' + i \, b_k'' ,$$  (5.9b)

one can interprete the characteristic functional as the Fourier transform of a probability density functional, which will be denoted by $P(\{a_k\})$ in spectral form. This leads us to

$$G(\{b_k\}) = \exp\{-\sum_k \lambda_k (b_k'^2 + b_k''^2)\} \tag{5.10}$$

$$= \langle \exp\{2i \sum_k (a_k' b_k' + a_k'' b_k'')\rangle$$

$$= \left( \prod_l \int_{-\infty}^{+\infty} da_l' \int_{-\infty}^{+\infty} da_l'' \right) P(\{a_k\}) \exp\{2i \sum_k (a_k' b_k' + a_k'' b_k'')\} \ .$$

Hence the probability density can be found by simply calculating the inverse Fourier transform

$$P(\{a_k\}) = \prod_k \frac{1}{\pi \lambda_k} \exp\left( -\frac{a_k^* a_k}{\lambda_k} \right) \ . \tag{5.11}$$

Note that, in order for these Fourier transforms to exist, one has to impose the condition

$$\text{Re } \lambda_k > 0 \qquad \text{(all } k) \tag{5.12}$$

on the function $g(x \mid x')$.

The continuous form of the probability density functional follows upon substitution of (5.6a, c) into (5.11)

$$P[\psi(x)] = N^{-1} \exp\{- \int dx \int dx' \ \bar{\psi}(x) \ g^{-1} (x \mid x') \ \psi(x')\} \ . \tag{5.13}$$

Here

$$g^{-1}(x \mid x') \equiv \sum_k \lambda_k^{-1} \chi_k(x) \bar{\chi}_k(x') \tag{5.14}$$

is the inverse of the kernel $g(x \mid x')$, and

$$N \equiv \int \exp\left\{- \int dx \int dx' \ \bar{\psi}(x) g^{-1} (x \mid x') \ \psi(x')\right\} d[\psi(x)] \tag{5.15}$$

symbolizes the normalization $\Pi_k(\pi \lambda_k)$ of the functional probability density.

Finally, we consider these results for the special case in which the asymmetric kernel $g(x \mid x')$ is given explicitly by the formulas (4.21). The right eigenvalue problem (5.5a) is satisfied by an eigenfunction of the form $(k \equiv n, l)$

$$\chi_k(\mathbf{r}, \tau) = U_n(\mathbf{r}) T_l(\tau) , \qquad (5.16)$$

where the $U_n (\mathbf{r})$ are defined by (2.7) and where

$$(1 - \zeta e^{-\beta E_n})^{-1} \int_0^\tau \exp\{-(\tau - \tau')E_n\} T_l(\tau') \, d\tau' + \zeta \exp(-\beta E_n)$$

$$\times (1 - \zeta e^{-\beta E_n})^{-1} \int_\tau^\beta \exp\{-(\tau - \tau')E_n\} T_l (\tau') \, d\tau' = \lambda_{n,l} T_l(\tau). \quad (5.17)$$

Differentiation of this equation with respect to $\tau$ gives

$$(1 - E_n \lambda_{n,l}) T_l (\tau) = \lambda_{n,l} \frac{d}{d\tau} T_l(\tau) , \qquad (5.18)$$

which has the solution

$$T_l(\tau) = \beta^{-1/2} \exp\left\{ - \left( E_n - \frac{1}{\lambda_{n,l}} \right) \tau \right\} . \qquad (5.19)$$

Substituting this general expression back into the integral equation (5.17) one finds

$$\frac{1}{\lambda_{n,l}} = E_n - \frac{\ln \zeta}{\beta} - \frac{2\pi l}{\beta} i, \qquad (l = 0, \pm 1, \pm 2, \ldots) . \qquad (5.20)$$

The corresponding right eigenfunction is

$$\chi_{n,l}(\mathbf{r}, \tau) = U_n(\mathbf{r}) \beta^{-1/2} \zeta^{-\tau/\beta} e^{-2\pi l \tau i/\beta} . \qquad (5.21)$$

In the same way the left eigenfunctions are found to be

$$\bar{\chi}_{n,l}(\mathbf{r}, \tau) = U_n^*(\mathbf{r}) \beta^{-1/2} \zeta^{+\tau/\beta} e^{+2\pi l \tau i/\beta} . \qquad (5.22)$$

Of course, the orthonormality condition (5.7) now reads

$$\int_\Omega d^3 r \int_0^\beta d\tau \, \bar{\chi}_{n,l}(\mathbf{r}, \tau) \chi_{n'l'} (\mathbf{r}, \tau) = \delta_{n,n'} \delta_{l,l'} . \qquad (5.23)$$

Our formula (5.20) also shows that the inverse kernel $g^{-1}(x \mid x')$ has the property that

$$\int g^{-1}(x \mid x') f(x') \, dx' = \left[ -\frac{\hbar^2}{2m} \Delta + W(\mathbf{r}) + \frac{\partial}{\partial \tau} \right] f(x) \qquad (5.24)$$

for any function $f(\mathbf{r}, \tau)$. Hence the probability density functional (5.13) can in this case also be written in the form

$$P[\psi(x)] = N^{-1} \exp\{-\int \bar{\psi}(x) \left[-\frac{\hbar^2}{2m}\Delta + W(\mathbf{r}) + \frac{\partial}{\partial \tau}\right] \psi(x) \, dx\} \ .$$

(5.25)

Writing

$$\psi(\mathbf{r}, \tau) = \zeta^{-\tau/\beta} \Psi(\mathbf{r}, \tau) \ ,$$ (5.26a)

$$\bar{\psi}(\mathbf{r}, \tau) = \zeta^{+\tau/\beta} \Psi^*(\mathbf{r}, \tau) \ ,$$ (5.26b)

one has the probability density functional

$$P[\Psi(\mathbf{r}, \tau)] = N^{-1} \exp\{-\int_\Omega d^3r \int_0^\beta d\tau \ \Psi^*(\mathbf{r}, \tau)$$

$$\times \left[-\frac{\hbar^2}{2m}\Delta + W(\mathbf{r}) - \mu + \frac{\partial}{\partial \tau}\right] \Psi(\mathbf{r}, \tau)\} \ ,$$ (5.27)

where

$$\mu \equiv \frac{\ln\zeta}{\beta}$$ (5.28)

and where the value of $N$ follows by normalization.

## 7.6  Bell's functional integral

In order to transform Wiegel's functional integral (4.17) into Bell's functional integral we consider the random functions $\psi(x)$ discussed in the previous section, with a covariance given by (4.21). Moreover, we have to restrict ourselves into interacting bosons. The case of interacting fermions will be briefly mentioned at the end of this section.

Consider the average

$$J[\phi(x)] = \langle \exp\{\int \bar{\psi}(x) \, \psi(x) \, \phi(x) \, dx\}\rangle_\psi \ ,$$ (6.1)

where the subscript $\psi$ indicates that the average has to be performed over the random functions $\psi(x)$, keeping $\phi(x)$ constant. Expanding the exponential and applying the decomposition theorem (5.2) to each term one finds

$$J[\phi(x)] = 1 + \sum_{N=1}^{\infty} \frac{1}{N!} \sum_P \int dx_1 \int dx_2 \ldots \int dx_N$$

$$\times \prod_{i=1}^{N} \langle \bar{\psi}(x_i) \, \psi(x_{Pi})\rangle_\psi \, \phi(x_{Pi})$$ (6.2)

Again we may decompose the permutations $P$ in cycles. The contribution of a permutation equals the product of the contributions of its cycles. In Eq. (2.9) we gave the number of permutations consisting of $n_s$ cycles of $s$ elements. Hence in the notation of Section 7.2

$$J[\phi(x)] = \sum_{N=0}^{\infty} \frac{1}{N!} \sum_{\{n_s\}}' \frac{N!}{\sum\limits_{s=1}^{\infty} (n_s! s^{n_s})}$$

$$\times \prod_{s=1}^{\infty} \left[ \int dx_1 \int dx_2 \ldots \int dx_s \prod_{j=1}^{s} \{g_W(x_{j+1} \mid x_j)\, \phi(x_{j+1})\} \right]^{n_s} , \quad (6.3)$$

where it is understood that $\phi(x_{s+1}) \equiv \phi(x_1)$ in the product over $j$. Performing the summations and products explicitly gives after some straightforward combinatorics:

$$J[\phi(x)] = \exp\left[ \sum_{s=1}^{\infty} \frac{1}{s} \int dx_1 \int dx_2 \ldots \int dx_s \prod_{j=1}^{s} \{\phi(x_{j+1})\, g_W(x_{j+1} \mid x_j)\} \right]$$

$$= \exp\{-\text{Tr} \ln(1 - \phi g_W)\} . \quad (6.4)$$

Now the right-hand side of this equation is identical to the integrand in the function integral (4.17). Hence one finds the repeated functional average

$$Z_B(z, \beta, \Omega) = Z_B^{(0)}(\zeta, \beta, \Omega) \langle\langle \exp\{ \int \bar\psi(x)\, \psi(x)\, \phi(x)\, dx\}\rangle_\psi\rangle_\phi , \quad (6.5)$$

where we have indicated that the averaging process over $\psi$ must be performed first, for fixed $\phi$, followed by an average over $\phi$. We shall assume that these two averaging processes may be interchanged, provided the volume $\Omega$ is finite. Using (VI.1.1), (VI.1.7) and (3.1) to perform the average over $\phi$ explicitly one finds Bell's functional integral

$$Z_B(z, \beta, \Omega) = Z_B^{(0)}(\zeta, \beta, \Omega) \langle \exp\{-\frac{1}{2} \int dx \int dx'\, \bar\psi(x)\, \psi(x)$$

$$\times V(x - x')\, \bar\psi(x')\, \psi(x')\}\rangle_\psi , \quad (6.6)$$

where

$$V(x - x') \equiv V(\mathbf{r} - \mathbf{r}')\, \delta(\tau - \tau') .$$

This result has been derived by Hirota [7], Bell [8], Edwards [9], Casher,

Lurié and Revzen [10], Jalickee and Wiegel [4] and Wiegel and Hijmans [5, 6] using a wide variety of methods. Also compare with Ref. IX.22.

In view of the considerable practical importance of Bell's integral we take a closer look at the meaning of the averaging symbol in (6.6). Introducing $\Psi(x)$ and its complex conjugate $\Psi^*(x)$ according to (5.26), combination of (6.6) with (5.27) gives the more explicit form

$$Z_B(z, \beta, \Omega) = Z_B^{(0)}(\zeta, \beta, \Omega) \prod_{n,l} \left\{ \frac{1}{\pi\lambda_{n,l}} \int_{-\infty}^{+\infty} da'_{n,l} \int_{-\infty}^{+\infty} da''_{n,l} \right\}$$

$$\times \exp\{-L[\Psi(x)]\} , \qquad (6.8a)$$

where

$$L[\Psi(x)] \equiv + \int \Psi^*(x) \left[ -\frac{\hbar^2}{2m}\Delta + W - \mu + \frac{\partial}{\partial\tau} \right] \Psi(x)dx$$

$$+ \frac{1}{2} \int dx \int dx' |\Psi(x)|^2 V(x - x') |\Psi(x')|^2 . \qquad (6.8b)$$

Now note the factor $Z_B^{(0)}(\zeta, \beta, \Omega)$ which occurs on the right-hand side, and which we calculated in Eqs. (2.13, 15)

$$Z_B^{(0)}(\zeta, \beta, \Omega) = \prod_n (1 - \zeta e^{-\beta E_n})^{-1} . \qquad (6.9)$$

This function has poles on the positive real $\zeta$-axis at $\zeta_n = e^{\beta E_n}$. However, these poles are spurious in the sense that they coincide with, and are cancelled by zeroes arising from the factor $\Pi_{n,l} \lambda_{n,l}^{-1}$. In order to make this more obvious we represent $Z_B^{(0)}$ itself by an infinite product of the form

$$Z_B^{(0)}(\zeta, \beta, \Omega) = \prod_{n,l} \frac{\lambda_{n,l}}{\nu_{n,l}} , \qquad (6.10)$$

where

$$\nu_{n,l}^{-1} = \xi_n - \frac{2\pi l}{\beta} i \qquad (6.11)$$

with $\xi_n$ still to be determined. Substituting (5.20) these equations give

$$Z_B^{(0)}(\zeta, \beta, \Omega) = \prod_{n,l} \frac{\beta\xi_n - 2\pi li}{\beta E_n - \beta\mu - 2\pi li} . \qquad (6.12)$$

We first perform the product over $l$. Multiplying the factors with $+l$ and $-l$ one finds for the right-hand side

$$Z_B^{(0)}(\zeta, \beta\, \Omega) = \prod_n \left\{ \prod_l \frac{\beta^2 \xi_n^2 + 4\pi^2 l^2}{\beta^2 (E_n - \mu)^2 + 4\pi^2 l^2} \right\}^{1/2}$$

$$= \prod_n \frac{1 - e^{-\beta\xi_n}}{1 - \zeta e^{-\beta E_n}}\, e^{-\beta(E_n - \mu - \xi_n)/2} , \qquad (6.13)$$

where the infinite product over $l$ has been evaluated by means of Eq. (1.431–2) of [11], and where we assumed

$$\xi_n > 0, \quad (E_n - \mu) > 0 . \qquad (6.14)$$

Comparing (6.13) with (6.9) we see that an infinite product representation of the form (6.10,11) is possible provided $\xi_n$ satisfies the condition

$$\exp\{\tfrac{1}{2}\beta(E_n - \mu)\} = 2 \sinh\tfrac{1}{2}\beta\xi_n . \qquad (6.15)$$

This equation gives exactly one positive real solution for $\xi_n$ for any real value of $\mu$. Combination of (6.8) and (6.10) thus leads to

$$Z_B(z, \beta, \Omega) = \int \exp\{-L[\Psi(x)]\}\, d[\Psi(x)], \qquad (6.16a)$$

where

$$\int d[\Psi(x)] \leftrightarrow \prod_{n,l} \left\{ \frac{1}{\pi\nu_{n,l}} \int_{-\infty}^{+\infty} da'_{n,l} \int_{-\infty}^{+\infty} da''_{n,l} \right\} . \qquad (6.16b)$$

The last formula demonstrates explicitly the absence of poles in the forefactor of the functional integral.

Moreover, we note that there is no reason to restrict the Bell functional integral to values of the chemical potential $\mu$ for which $E_0 - \mu > 0$. This condition was, of course, required to guarantee the convergence of the integral of (5.27) over the whole function space. But with a fourth-order term present in (6.8b) and (6.16a) the functional integral for the partition function will converge also if $E_n - \mu < 0$ for some $n$, provided

$$\int V(\mathbf{r})\, d^3r > 0. \qquad (6.17)$$

We shall assume that this is always the case in the applications which we are going to make.

Finally, the reader should note that Bell's functional integral holds for bosons only. Its counterpart for fermions can be written down only by interpreting the $\psi(x)$ as non-commuting matrices. This, of course, spoils the beauty of the whole path-integral approach to quantum fluids, which works with commuting quantities like complex numbers throughout. Apart from these aesthetic considerations a practical reason to mistrust the "functional integrals over non-commuting variables" which arise for fermions is the difficulty of finding sensible approximation schemes and of doing actual calculations.

## References

[1]   A.J.F. Siegert, *Physica* **26** (1960) S30.
[2]   R.L. Stratonovich, *Sov. Phys. Doklady* **2** (1958) 416.
[3]   S.F. Edwards, *Phil. Mag.* **4** (1959) 1171.
[4]   J.B. Jalickee and F.W. Wiegel, *Physica* **48** (1970) 589.
[5]   F.W. Wiegel and J. Hijmans, *Proc. Acad. Sci.* Amsterdam **B77** (1974) 177.
[6]   F.W. Wiegel and J. Hijmans, *Proc. Acad. Sci.* Amsterdam **B77** (1974) 189.
[7]   R. Hirota, thesis, Northwestern University (1961), unpublished.
[8]   J.S. Bell, in *Lectures on the Many-Body Problem*, E.R. Caianiello, ed. (Academic Press, New York, 1962).
[9]   S.F. Edwards, in *Analysis in Function Space*, W.T. Martin and I. Segal, eds. (M.I.T. Press, Cambridge, MA., 1964) p. 167.
[10]  A. Casher, D. Lurié and M. Revzen, *J. Math. Phys.* **9** (1968) 1312.
[11]  I.S. Gradshteyn and I.M. Ryzhik, *Tables of Integrals, Series and Products* (Academic Press, New York, 1965).

# VIII. APPROXIMATE THEORIES OF THE INTERACTING BOSE FLUID AND THE VORTEX-RING MODEL FOR THE LAMBDA TRANSITION

After the rather formal developments of Chapter VII we show in this chapter that application of the saddle-point method to the functional integral representations of the partition function actually does lead to interesting physics. These calculations are most straightforward when based on Bell's representation (VII.6.16) with the "Lagrangian" $L$ given by (VII.6.8b). This approach will be followed here. For an older theory of the $\lambda$ transition in $^4$He, based on the Feynman path integral (VII.1.9), the reader should consult the literature quoted before [I.20–I.28] as well as a classic paper by Cohen and Feynman [25].

## 8.1 Properties of the stationary points

The integrand of (VII.6.16), considered as a functional of $\Psi$, defines a surface in an infinite dimensional space. The stationary points of this surface are functions (denoted by $\hat{\Psi}$ in this chapter) for which the variations of $L$ with respect to $\Psi$ and $\Psi^*$ vanish. This leads to two equations

$$\left[ -\frac{\hbar^2}{2m}\Delta + W - \mu + \frac{\partial}{\partial\tau} + \int V(x - x') \, |\hat{\Psi}(x')|^2 \, dx' \right] \hat{\Psi}(x) = 0 \; ,$$

$$(1.1a)$$

$$\left[ -\frac{\hbar^2}{2m}\Delta + W - \mu - \frac{\partial}{\partial\tau} + \int V(x - x') \, |\hat{\Psi}(x')|^2 \, dx' \right] \hat{\Psi}^*(x) = 0 \; .$$

$$(1.1b)$$

In deriving these equations we used the representation

$$\Psi(x) = \sum_{n,l} \beta^{-1/2} \, U_n(\mathbf{r}) \, e^{-2\pi l \tau i/\beta} \; , \qquad (1.2)$$

which follows from (VII.5.6, 21, 26), to set certain boundary terms arising in the process of partial integration equal to zero. Taking the complex conjugate of (1.1b) and comparing with (1.1a) one finds $\partial\hat{\Psi}/\partial\tau = 0$, which implies that the stationary points of the functional surface are functions $\hat{\Psi}(\mathbf{r})$ which do not depend on the auxiliary and rather unphysical variable $\tau$. Consequently both (1.1a, b) reduce to the same equation

$$\left[ -\frac{\hbar^2}{2m}\Delta + W(\mathbf{r}) - \mu + \int V(\mathbf{r} - \mathbf{r}') \, | \, \hat{\Psi}(\mathbf{r}') \, |^2 \, d^3r' \right] \hat{\Psi}(\mathbf{r}) = 0 \; . \quad (1.3)$$

The expansion (1.2) shows that this equation must be solved subject to the boundary conditon that $\hat{\Psi} = 0$ on a hard wall, and $\hat{\Psi} \to 0$ for $r \to \infty$ in all directions in which there is no hard wall.

In many applications one is interested in solutions which do not vary appreciably over distances of the order of the range of the pair potential $V(\mathbf{r} - \mathbf{r}')$. For such solutions the stationarity equation has the local form

$$\left[ -\frac{\hbar^2}{2m}\Delta + W(\mathbf{r}) - \mu + \bar{V}(0) \, | \, \hat{\Psi} \, |^2 \right] \hat{\Psi} = 0 \; , \qquad (1.4)$$

where all arguments are $\mathbf{r}$ and where $\bar{V}(0)$ denotes the $\mathbf{k} = 0$ component of the Fourier transform of the pair potential

$$\bar{V}(\mathbf{k}) = \int V(\mathbf{r}) \exp(i\mathbf{k}\cdot\mathbf{r}) \, d^3r \; . \qquad (1.5)$$

In the rest of this chapter we assume $\bar{V}(0) > 0$, corresponding to a predominantly repulsive potential. We shall also usually take $W(\mathbf{r}) = 0$, corresponding to the absence of external forces. The stationary equation, which now reads

$$-\frac{\hbar^2}{2m}\triangle\hat{\Psi} - \mu\hat{\Psi} + \bar{V}(\mathbf{0})\,|\hat{\Psi}|^2\,\hat{\Psi} = 0 \qquad (1.6)$$

has the form of the Gross-Pitaevskii equation [1]. Note that (1.6) is *not* a Landau-Ginzburg equation, because a Landau-Ginzburg equation is always assumed ad hoc in a phenomenological theory, whereas (1.6), like the Gross-Pitaevskii equation, results from a microscopic theory after an explicit derivation. It is unfortunate that this confusion of terminology still persists in some of the literature.

In order to further analyze the solutions of (1.6) we decompose the number $\hat{\Psi}$ into a modulus $A$ and a phase $\phi$ which should, of course, not be confused with the Gaussian random field $\phi$ which played a role in earlier chapters

$$\hat{\Psi}(\mathbf{r}) = A(\mathbf{r})\,\exp\{i\phi(\mathbf{r})\}\ . \qquad (1.7)$$

Substituting into (1.6) and separating the real and imaginary parts one finds

$$2(\nabla A)\cdot(\nabla\phi) + A\triangle\phi = 0\ , \qquad (1.8a)$$

$$-\frac{\hbar^2}{2m}\triangle A + \frac{\hbar^2}{2m}A(\nabla\phi)^2 - \mu A + \bar{V}(\mathbf{0})A^3 = 0\ . \qquad (1.8b)$$

The following theorem shows the special role which is played by the point $\mu = 0$ as far as the uniqueness or multiplicity of the solutions of the stationarity equation is concerned:

**Theorem:** If $\mu < 0$, then $\hat{\Psi}(\mathbf{r}) = 0$ is the only solution of (1.6). If $\mu > 0$, many solutions of (1.6) exist, all satisfying the inequality

$$A(\mathbf{r}) \le \{\mu/\bar{V}(\mathbf{0})\}^{1/2}\ . \qquad (1.9)$$

In order to prove this let us assume that, although $\mu$ is negative a non-zero solution exists. There would exist, therefore, a point $\mathbf{r}_0$ in which $A(\mathbf{r}_0) > 0$. As the solutions must be continuous the inequality $A(\mathbf{r}) > 0$ must hold in a finite region around $r_0$. Let $\Omega_1$ denote the largest such region. Now apply Gauss' theorem in the form

$$\frac{\hbar^2}{2m}\int_{\Omega_1}\triangle A\ d^3r = \frac{\hbar^2}{2m}\int_{S_1}\int\nabla A\cdot d\mathbf{S}\ , \qquad (1.10)$$

where the surface integral is extended over the boundary $S_1$ of $\Omega_1$. Obviously $A = 0$ on $S_1$, but $A > 0$ inside $S_1$. Therefore $\nabla A$ points inwards and the right-hand-side of (1.10) would be $\leq 0$. The left-hand-side, however, can be written with (1.8b) as

$$\frac{\hbar^2}{2m} \int_{\Omega_l} \triangle A \, d^3r = \int_{\Omega_1} \left[ \frac{\hbar^2}{2m} A (\nabla\phi)^2 - \mu A + \bar{V}(0) \, A^3 \right] d^3r \, , \qquad (1.11)$$

which is $> 0$, leading to a contradiction. Thus no solutions can exist which are non-zero at some point in the volume. The second part of our theorem can be proved in an analogous way if one assumes $\mu > 0$ and a solution which violated (1.9). Define $\Omega_1$ as the largest region inside which $A > \{\mu/\bar{V}(0)\}^{1/2}$. Using (1.10) and (1.11) one again finds a contradiction.

For the case $\mu > 0$ one can bring the stationarity equation in a dimensionless form by introducing a dimensionless position vector $\rho$ and a dimensionless field $\Phi$ through the relations

$$\rho \equiv (2m\mu/\hbar^2)^{1/2} \, \mathbf{r} \, , \qquad (1.12)$$

$$\Phi \equiv (\bar{V}(0)/\mu)^{1/2} \, \hat{\Psi} \, . \qquad (1.13)$$

Now the stationarity equation, in the case $W(\mathbf{r}) = 0$, gets the dimensionless form

$$-\triangle_\rho \Phi - \Phi + |\Phi|^2 \, \Phi = 0 \, , \qquad (1.14)$$

where the Laplacian is with respect to the components of $\rho$. This transformation shows that the solutions of the stationarity equation, in the absence of an external potential, will vary in space on a scale which is determined by the "healing length"

$$l \equiv (\hbar^2/2m\mu)^{1/2} \, . \qquad (1.15)$$

Consequently, the replacement of (1.3) by (1.4) is permitted provided the order of magnitude estimate

$$(\text{range of } V(\mathbf{r})) \ll l \qquad (1.16)$$

holds true; this will of course be the case for sufficiently small, positive values of $\mu$.

## 8.2  The method of the maximum term and the effective field theory

For any solution of the stationarity equation one can determine its contribution to the functional integral by substitution into (VII.6.8b) and elimination of the $\Psi^* \triangle \Psi$ term with the use of (1.4). This gives, for an arbitrary external potential $W(\mathbf{r})$

$$\exp\{-L[\hat{\Psi}(\mathbf{r})]\} = \exp\{+\frac{1}{2}\beta\bar{V}(\mathbf{0})\int_{\Omega} A^4(\mathbf{r})\,d^3\mathbf{r}\} . \qquad (2.1)$$

Specializing again to the case in which no external potential is present, $W = 0$, and using our theorem one finds that for $\mu > 0$ the absolute maximum in function space is attained for the field

$$\hat{\Psi}_0(\mathbf{r}) = \left\{\frac{\mu}{V(\mathbf{0})}\right\}^{1/2} \exp(i\phi_0) , \qquad (\mu > 0) , \qquad (2.2)$$

where $\phi_0$ is a constant, but arbitrary phase. Close to the walls $\hat{\Psi}_0$ will of course drop from this value to zero.

In a more geometric language one can picture the integrand $\exp(-L)$ as a surface over function space (cf. Ref. 2). For $\mu < 0$ this functional surface resembles a hill with a peak at the origin of $\Psi$ space. For $\mu > 0$ the region of the functional surface near the absolute maximum resembles the ridge of a volcano, due to the phase degeneracy in (2.2).

The method of the maximum term simply amounts to the approximation

$$Z \cong \exp\{-L[\hat{\Psi}_0]\} . \qquad (2.3)$$

The grand canonical pressure is

$$\beta p(\mu, \beta) = \lim_{\Omega \to \infty} \frac{1}{\Omega} \ln Z$$

$$= 0 \qquad (\mu < 0) ,$$

$$= \frac{\beta\mu^2}{2\bar{V}(\mathbf{0})} \qquad (\mu > 0) . \qquad (2.4)$$

The grand canonical density is

$$\rho(\mu, \beta) = \frac{\partial p(\mu, \beta)}{\partial \mu}$$

$$= 0 \qquad (\mu < 0) ,$$

$$= \frac{\mu}{\bar{V}(\mathbf{0})} \qquad (\mu > 0) . \qquad (2.5)$$

Hence the isotherm (pressure as a function of density and temperature) is found by elimination of $\mu$

$$p(\rho, T) = \frac{1}{2} \tilde{V}(\mathbf{0}) \rho^2 \ . \tag{2.6}$$

This approximation, which can be called the effective field theory because it tries to describe the system with a single, constant field of the optimal strength (2.2), is obviously too crude to give anything of physical interest. We shall therefore hasten to improve the calculation in the following sections.

### 8.3 The quadratic approximation and the Bogoliubov theory

The failure of the effective field theory is due to the fact that only the contribution of the maximum of the integrand of Eq. (VII.6.16) was taken into account. The contribution of the *vicinity* of the maximum was neglected altogether. In this section we calculate this contribution in the quadratic approximation, following Ref. [3]. Hence this section does for interacting bosons what Section 1.6 did for Brownian movement in a field with a quadratic particle-annihilation strength, and what Section 2.3 did for Brownian particles subject to external forces. It turns out that one recovers in this way the results of a much older theory of the interacting Bose fluid, due to Bogoliubov.

As a preliminary we investigate the curvatures of the functional surface in a point of the ridge, for the case $\mu > 0$, $W = 0$. As we will be interested in the thermodynamic limit we may suppose the volume to have the shape of a cube with sides of length $\Omega^{1/3}$, and we may neglect the surface terms which are introduced if we impose periodic boundary conditions on the functions $U(\mathbf{r})$ of Eq. (VII.2.7) instead of hard wall boundary conditions. We shall thus work with

$$U_{\mathbf{k}}(\mathbf{r}) = \Omega^{-1/2} \exp(i\mathbf{k} \cdot \mathbf{r}) \ , \tag{3.1}$$

where the components of $\mathbf{k}$ equal $2\pi\Omega^{-1/3}$ times an integer. Substituting (3.1) into (1.2) one finds

$$\Psi(x) = \sum_{\mathbf{k},l} a_{\mathbf{k},l} (\beta\Omega)^{-1/2} \exp(i\mathbf{k} \cdot \mathbf{r} - 2\pi i l\tau/\beta) \ . \tag{3.2}$$

When this expansion is substituted into (VII.6.8b) the Lagrangian becomes, after a straightforward calculation,

$$L[\{a_\sigma\}] = \sum_\sigma \left( \varepsilon_{\mathbf{k}} - \mu - \frac{2\pi l}{\beta} i \right) |a_\sigma|^2$$

$$+ \frac{1}{2\beta\Omega} \sideset{}{'}\sum_{\sigma_1,\sigma_2,\sigma_3,\sigma_4} a_{\sigma_1}^* a_{\sigma_2} a_{\sigma_3}^* a_{\sigma_4} \tilde{V}(k_4 - k_3) \ . \tag{3.3}$$

Here $\sigma \equiv \mathbf{k}, l$, the prime indicates the constraint

$$\sigma_1 + \sigma_3 = \sigma_2 + \sigma_4 \tag{3.4}$$

and

$$\varepsilon_{\mathbf{k}} \equiv \hbar^2\mathbf{k}^2/2m \tag{3.5}$$

are the free-boson energy eigenvalues.

In the vicinity of the ridge we put

$$\Psi(x) = \left[\left\{\frac{\mu}{\tilde{V}(0)}\right\}^{1/2} + \rho_0\right] \exp(i\phi_0)$$

$$+ \sum_{\mathbf{k},l}' a_{\mathbf{k},\,l}\,(\beta\Omega)^{-1/2}\exp(i\mathbf{k}\cdot\mathbf{r} - 2\pi i l\tau/\beta)\ , \tag{3.6}$$

where the prime means omission of the term with $\mathbf{k} = \mathbf{0}, l = 0$ and where

$$-\left\{\frac{\mu}{\tilde{V}(0)}\right\}^{1/2} < \rho_0 < \infty\ . \tag{3.7}$$

Near the ridge the quantities $\rho_0$ and $(\beta\Omega)^{-1/2}\,a_{\mathbf{k},l}$ are small as compared to $\{\mu/\tilde{V}(0)\}^{1/2}$. If we substitute (3.6) into (3.3) and omit all terms of higher than second order in these small quantities, we find after an inelegant calculation

$$L[\{a_\sigma\}] \cong -\frac{\mu^2\beta\Omega}{2\tilde{V}(0)} + 2\mu\rho_0^2\,\beta\Omega$$

$$+ \sum_\sigma' \left[\left(\varepsilon_{\mathbf{k}} + \mu\frac{\tilde{V}(\mathbf{k})}{\tilde{V}(0)} - \frac{2\pi l}{\beta}i\right)a_\sigma^* a_\sigma\right.$$

$$\left. + \frac{\mu\tilde{V}(\mathbf{k})}{2\tilde{V}(0)}\,e^{2i\phi_0}\,a_\sigma^*\,a_{-\sigma}^* + \frac{\mu\tilde{V}(\mathbf{k})}{2\tilde{V}(0)}\,e^{-2i\phi_0}\,a_\sigma\,a_{-\sigma}\right]\ . \tag{3.8}$$

This shows that the surface is flat if one moves through a point of the ridge in the direction of increasing phase angle $\phi_0$. Also one should note that the second derivative with respect to any of the quantities $\rho_0$ or $(\beta\Omega)^{-1/2}\,a_\sigma$ with $\sigma \neq 0$ is of the order $\mu\beta\Omega$. This implies that the ridge is very sharp if one moves in those directions.

In order to find the curvatures of the functional surface in the directions of the $a_\sigma$ with $\sigma \neq 0$ we have to diagonalize the expression $L_\sigma + L_{-\sigma}$ where

$$L_\sigma \equiv \left( \varepsilon_\mathbf{k} + \mu \frac{\tilde{V}(\mathbf{k})}{\tilde{V}(\mathbf{0})} - \frac{2\pi l}{\beta} i \right) a_\sigma^* a_\sigma$$

$$+ \frac{\mu \tilde{V}(\mathbf{k})}{2\tilde{V}(\mathbf{0})} \left( e^{+2i\phi_0} a_\sigma^* a_{-\sigma}^* + e^{-2i\phi_0} a_\sigma a_{-\sigma} \right) . \tag{3.9}$$

First rotate the coordinate axis in the 2-dimensional complex $(a_\sigma, a_{-\sigma})$ space over an angle $-\phi_0$ by putting

$$a_\sigma = b_\sigma e^{i\phi_0} , \qquad (\sigma \neq 0) . \tag{3.10}$$

This transformation preserves the area, i.e., in Eq. VII.6.16b the element $da_\sigma' \, da_\sigma''$ will transform into $db_\sigma' \, db_\sigma''$. In the complex $(b_\sigma, b_{-\sigma})$ space one now has to diagonalize

$$L_\sigma + L_{-\sigma} = \gamma_\sigma b_\sigma^* b_\sigma + \gamma_{-\sigma} b_{-\sigma}^* b_{-\sigma}$$

$$+ \delta_\sigma (b_\sigma^* b_{-\sigma}^* + b_\sigma b_{-\sigma}) , \tag{3.11}$$

where

$$\gamma_\sigma = \varepsilon_\mathbf{k} + \frac{\tilde{V}(\mathbf{k})}{\tilde{V}(\mathbf{0})} - \frac{2\pi l}{\beta} i , \tag{3.12}$$

$$\delta_\sigma = \mu \frac{\tilde{V}(\mathbf{k})}{\tilde{V}(\mathbf{0})} \tag{3.13}$$

This is most easily done if one uses a vector notation. It is easily seen that

$$L_\sigma + L_{-\sigma} = (\xi_\sigma, \Lambda_\sigma \xi_\sigma) , \tag{3.14}$$

where $\xi_\sigma$ is the complex 2-dimensional vector

$$\xi_\sigma = \begin{pmatrix} b_\sigma \\ b_{-\sigma}^* \end{pmatrix} \tag{3.15}$$

and $\Lambda_\sigma$ the complex $2 \times 2$ matrix

$$\Lambda_\sigma = \begin{vmatrix} \gamma_\sigma & \delta_\sigma \\ \delta_\sigma & \gamma_{-\sigma} \end{vmatrix} . \tag{3.16}$$

The two eigenvalues $\lambda_{\sigma, \pm}$ of this matrix are found to be equal to

$$\lambda_{\sigma, \pm} = \varepsilon_\mathbf{k} + \mu \frac{\tilde{V}(\mathbf{k})}{\tilde{V}(\mathbf{0})} \pm \left\{ \mu^2 \frac{\tilde{V}(\mathbf{k})^2}{\tilde{V}(\mathbf{0})^2} - \frac{4\pi^2 l^2}{\beta^2} \right\}^{1/2} . \tag{3.17}$$

These eigenvalues are real for $4\pi^2 l^2 \le \mu^2 \beta^2 \tilde{V}^2(\mathbf{k})/\tilde{V}^2(\mathbf{0})$; otherwise they are complex numbers with a real part equal to $\varepsilon_\mathbf{k} + \mu \tilde{V}(\mathbf{k})/\tilde{V}(\mathbf{0})$. Note that

$$\lambda_{\sigma,+}\, \lambda_{\sigma,-} = E_\mathbf{k}^2 + \frac{4\pi^2 l^2}{\beta^2} \,,$$

$$E_\mathbf{k} = \sqrt{\varepsilon_\mathbf{k}\left(\varepsilon_\mathbf{k} + 2\mu\, \frac{\tilde{V}(\mathbf{k})}{\tilde{V}(\mathbf{0})}\right)} \,. \tag{3.18}$$

Of course, the curvature of the functional surface in a point of the ridge is essentially determined by the quantities $\beta\Omega\, \lambda_{\sigma,+}$ and $\beta\Omega\, \lambda_{\sigma,-}$.

In order to calculate the partition function in the quadratic approximation we return to Eq. (VII.6.16) and notice that the 4-dimensional integral over $a_\sigma'$, $a_\sigma''$, $a_{-\sigma}'$, $a_{-\sigma}''$ (where $\sigma \ne 0$) equals

$$\int d^2 a_\sigma \int d^2 a_{-\sigma}\, \exp(-L_\sigma - L_{-\sigma}) = \frac{\pi^2}{\lambda_{\sigma,+}\, \lambda_{\sigma,-}} = \frac{\pi^2 \beta^2}{\beta^2 E_\mathbf{k}^2 + 4\pi^2 l^2} \,, (\sigma \ne 0) \,. \tag{3.19}$$

There is one such factor in (VII.6.16a) for every pair $+\sigma$, $-\sigma$ with $\sigma \ne 0$. Moreover, for such a pair (VII.6.16b) shows that there is a normalization factor equal to

$$(\pi^2 \nu_\sigma \nu_{-\sigma})^{-1} = \pi^{-2}\left(\xi_\mathbf{k}^2 + \frac{4\pi^2 l^2}{\beta^2}\right) \,, \tag{3.20}$$

where we used (VII.6.11,15) to show that $\xi_\mathbf{k} = \xi_{-\mathbf{k}}$. Finally, the integration over $a_0$ leads to a factor

$$\exp\left\{\frac{\mu^2 \beta\Omega}{2\tilde{V}(\mathbf{0})}\right\}$$

due to the first term in (3.8), times another factor which does not contribute to the thermodynamic functions in the limit $\Omega \to \infty$. Collecting these factors one finds in the quadratic approximation for the grand canonical partition function

$$Z(z, \beta, \Omega) \cong \exp\left\{\frac{\mu^2 \beta\Omega}{2\tilde{V}(\mathbf{0})}\right\} \prod_\sigma{}' \frac{\beta^2 \xi_\mathbf{k}^2 + 4\pi^2 l^2}{\beta^2 E_\mathbf{k}^2 + 4\pi^2 l^2} \,, \tag{3.21}$$

where the prime on the reproduct sign indicates that every pair $+\sigma$, $-\sigma$ with $\sigma \ne 0$ contributes only one factor. The infinite product can be evaluated with the help of the formula

$$\prod_{l=-\infty}^{+\infty} \frac{R^2 + 4\pi^2 l^2}{S^2 + 4\pi^2 l^2} = \frac{(1 - e^R)\,(1 - e^{-R})}{(1 - e^S)\,(1 - e^{-S})} \tag{3.22}$$

a proof of which can be found in [3]. In order to do this one simply notes that one gets the square of the product in (3.21) if one permits one factor to occur for each $\sigma$, but still omits $\sigma = 0$. Using this, and (VII.6.15), it is straightforward to show that in the thermodynamic limit

$$Z(z, \beta, \Omega) \cong \exp\left\{\frac{\mu^2 \beta \Omega}{2\tilde{V}(0)}\right\} \prod_{k \neq 0} \frac{e^{\beta(\varepsilon_k - \mu - E_k)/2}}{1 - e^{-\beta E_k}} . \tag{3.23}$$

Using (2.4) the grand canonical pressure is found to be given by

$$\beta p(\mu, \beta) = \frac{\beta \mu^2}{2\tilde{V}(0)} + \frac{1}{2(2\pi)^3} \int \beta(\varepsilon_k - \mu - E_k) \, d^3k$$

$$- \frac{1}{(2\pi)^3} \int \ln(1 - e^{-\beta E_k}) \, d^3k . \tag{3.24}$$

Here we used the standard recipe

$$\sum_k \to \frac{\Omega}{(2\pi)^3} \int d^3k \tag{3.25}$$

to replace a sum over one-particle states in a cube by a continuous integral, in the limit of a very large cube. The derivation here follows Ref. 2; the corresponding derivation of the grand canonical density can be found in Ref. 3.

The various terms in (3.24) have an interesting physical interpretation. The first term on the right-hand side is identical to the effective field result (2.4), i.e., it comes from a uniform mean field (the "condensate") in which the bosons move independently. The second term has to be made finite by terminating the integration at some large value $k_c$ of $|\mathbf{k}|$; usually one takes

$$k_c = 2\pi/(\text{range of } V(\mathbf{r})) . \tag{3.26}$$

With this cut-off the second term in (3.24) is finite and of no physical interest. The main term is the last one, to be denoted by

$$\beta p_B(\mu, \beta) = \frac{-1}{(2\pi)^3} \int \ln(1 - e^{-\beta E_k}) \, d^3k . \tag{3.27}$$

Comparing this expression with our earlier result (VII.2.16) it is seen that $p_B$ is the grand canonical pressure of a gas of *non-interacting* bosons with an energy spectrum given by $E_k$. These fictitious particles are called the quasiparticles of the system. The "quasi activity" of this gas of non-interacting quasiparticles is equal to unity. For small values of $|\mathbf{k}|$, hence for long wave lengths of the quasiparticles, their energy is given by the small $k$ approximation to (3.18)

$$E_{\mathbf{k}} \cong (2\mu \ \varepsilon_{\mathbf{k}})^{1/2} = \alpha \mid \mathbf{k} \mid \ ,$$ (3.28a)

where

$$\alpha \equiv (\mu \hbar^2/m)^{1/2} \ .$$ (3.28b)

Hence a plot of $E_k$ versus $k$ starts out linearly near the origin. For large values of $k$ one has $E_{\mathbf{k}} \cong \varepsilon_{\mathbf{k}}$. This linear behavior of $E_{\mathbf{k}}$ for small $|\mathbf{k}|$ can be shown to imply that the quasiparticles are quantized sound waves, also called phonons.

These results are identical to the results of the theory of Bogoliubov, as extended by Glassgold, Kaufman and Watson [4] to temperatures above the absolute zero, but far below the transition temperature. This shows that Bogoliubov's operator method for the Bose fluid is equivalent to the quadratic approximation of Bell's functional integral. We found that for $\mu > 0$ the interacting Bose gas essentially behaves as a uniform condensate plus a collection of phonon-like quasiparticles. We also found that the energies of these quasiparticles are related, through Eqs. (3.17, 18), to the curvatures of the functional surface at the ridge, in the directions of the principal axes.

The interested reader who would like to pursue these matters further, should also study the works of Langer, Fisher, Reppy and Rice in Refs. 5–8, which is somewhat related to the contents of this section.

## 8.4  Vortex-ring model for the lambda transition

In this and the next section we take the third step in our program to evaluate Bell's functional integral (VII.6.16): we try to take into account not just the contribution of the absolute maximum (Section 8.2), or the contribution of the *vicinity* of the absolute maximum (Section 8.3), but we try to find *all* maxima of the functional and sum their various contributions to the partition function. Hence we now try to do for interacting bosons what Section 2.4 did for Brownian particles subject to external forces, or what Section 6.3 did for the model of Kac, Uhlenbeck and Hemmer in the case $\gamma > 0$. We shall follow Ref. 9.

In order to find the proper point of view to start this calculation one should go back to the natural phenomena themselves and realize that the interacting Bose fluid is meant to be a model for $^4$He. One of the many striking phenomena which occurs in liquid helium is a logarithmic singularity in its specific heat, the $\lambda$-transition [10]. The resemblance between this singularity and the singularities associated with various order-disorder transitions suggests that the $\lambda$-transition is due to some type of collective behavior. The nature of this collective phenomenon is at present still somewhat mysterious.

In order to find the stationary points of the integrand in Bell's functional integral we return to the stationarity equation in its amplitude-and-phase-separated form (1.8a, b). Multiplying the first of these equations with $A$ we get

$$\text{div } (A^2 \nabla \phi) = 0 \ . \tag{4.1}$$

We shall interprete this equation as describing the flow of some "field" associated with the liquid helium system. Now the dimension of $A$ is $[\text{length}]^{-3/2}$, and $\phi$ is dimensionless. Hence $A^2$ has the dimension of an inverse volume. This naturally suggests the following identifications:
(a)   number density of the particles associated with the field, $A^2(\mathbf{r})$;
(b)   mass density of the field: $mA^2(\mathbf{r})$;
(c)   flow velocity of the field: $(\hbar/m) \nabla \phi$;
(d)   momentum density of the field: $\hbar A^2(\mathbf{r}) \nabla \phi$.

Now Eq. (4.1) tells us that the momentum density of the field is a conserved quantity. As a result of this "hydrodynamic" interpretation every solution of the stationarity equation can be identified with a certain flow pattern of the field.

Moreover, we note the boundary condition $A = 0$ on a hard wall. This implies, through Eq. (1.8a), the $\nabla \phi$ will be parallel to the wall. Hence the velocity of the field at a hard wall must be parallel to the wall, which means that the field "slips" at a wall, just like an ideal fluid. *It is, therefore, tempting to identify this field with the microscopic flow pattern of the superfluid component of the liquid helium system.*

The fact that the field is in motion (for solutions with a non-constant phase) does not contradict the fact that we are concerned with the properties of the Bose fluid in thermal equilibrium. The flow of the field should be thought of as occurring on a microscopic scale, given by the healing length (1.15), and is the result of the heat motion of the bosons. What Bell's functional integral tells us is that the partition function of the system is a sum of contributions from the very many ways in which the fluid, which is at rest on a macroscopic scale, can be stirred on a microscopic scale!

The identification of the flow velocity $\mathbf{v}$ with $(\hbar/m) \nabla \phi$ implies for the circulation of the field around a closed curve $C$

$$\int_C \mathbf{v} \cdot \mathbf{d}l = \frac{\hbar}{m} \int_C \nabla \phi \cdot \mathbf{d}\mathbf{l} = \frac{\hbar}{m} (\phi_2 - \phi_1) \ , \tag{4.2}$$

where $\phi_2 - \phi_1$ denotes the difference of the values of the phase, in a specific point of $C$, depending on the direction of approach of this point along the curve $C$. As the solution $\hat{\Psi}$ of (1.6) must be single valued, $\phi_2 - \phi_1$ must be equal to an integer $(n)$ times $2\pi$, leading to

$$\oint_C \mathbf{v} \cdot d\mathbf{l} = n\,\frac{h}{m} \,, \qquad (n = 0, \pm 1, \pm 2, \ldots) \,. \tag{4.3}$$

This is the well-known quantization of the circulation in superfluid flow in units of $h/m$, first conjectured by Onsager [11].

In order to obtain more information about the general structure of the solutions of the stationarity equation (1.6) we classify the solutions according to the dimension of the set of points where they vanish. Accordingly one should consider solutions with nodal points, solutions with nodal lines, solutions with nodal surfaces, and solutions with some mixture of nodal points, lines and surfaces. In this chapter we shall assume (ad hoc!) that the nodal line solutions are the dominating ones for helium close to the $\lambda$-transition, and that all other solutions can be neglected in that regime.

A detailed study of the nodal-like solutions is simplest in the case in which the nodal line is a single, infinite straight line. Transforming (1.6) to cylindrical coordinates $(z, r, \theta)$ with the $z$-axis along the nodal line (the coordinate $z$ should of course not be confused with the fugacity $z$), one finds

$$-\frac{\hbar^2}{2m}\left(\frac{\partial^2}{\partial r^2} + \frac{1}{r}\frac{\partial}{\partial r} + \frac{1}{r^2}\frac{\partial^2}{\partial \theta^2} + \frac{\partial^2}{\partial z^2}\right)\hat{\Psi} - \mu\hat{\Psi} + \bar{V}(0)\,|\hat{\Psi}|^2\hat{\Psi} = 0 \,.$$
$$\tag{4.4}$$

Trying a $z$-independent solution of the form

$$\hat{\Psi}(r, \theta) = f(r)\,\exp\{i(\theta_0 + n\theta)\} \,, \tag{4.5}$$

where $f(r)$ is real and $n$ an integer, one finds the equation

$$\frac{d^2 f}{dr^2} + \frac{1}{r}\frac{df}{dr} - \frac{n^2}{r^2}f + \frac{2m\mu}{\hbar^2}f - \frac{2m}{\hbar^2}\bar{V}(0)\,f^3 = 0 \,. \tag{4.6}$$

The boundary conditions are

$$f(0) = 0 \,, \qquad f(\infty) = \{\mu/\bar{V}(0)\}^{1/2} \,, \tag{4.7}$$

(cf. Eq. (1.9)). The solution of (4.6, 7) has been considered extensively in the literature; the reader might want to consult the review papers by

Gross [1] and by Fetter [12]. The velocity of the flow has no components in the $r-$ or $z$-directions; the component $v_\theta$ in the direction of increasing values of $\theta$ is

$$v_\theta(r) = \frac{n\hbar}{mr} , \qquad (n = 0, \pm 1, \pm 2, \ldots) . \qquad (4.8)$$

This is the velocity field of a classical vortex line, with circulation $nh/m$. We conclude that the flow patterns corresponding to the nodal-line solutions are quantized vortex lines.

Vortex lines in $^4$He have been observed, for example, in the experiments of Andronikashvili and Mamaladze [13] and Reppy [8] and also form the basis of theoretical work by Langer and Fisher [6]. The idea that the $\lambda$-transition is caused by the sudden appearance of very many extremely long vortex lines (an idea to be worked out in the rest of this chapter) underlines the work of Byckling [14], the author [9] and of Popov [15]. This idea has later emerged in quantum field theory, especially in the work of Rasetti and Regge [16], Banks, Myerson and Kogut [17], Kogut [18], Samuel [19, 20] and Hasenfratz [21, 22]. The dynamics of quantum vortices has been studied especially by Creswick and Morrison [23]; also compare with the dissertation of Creswick [24].

Equations (4.4–8) give results for an infinitely long, straight vortex line. We have also argued that all solutions of the stationarity equation vary on a spatial scale $l$ given by (1.15). From this observation one concludes that (1.6) also has solutions which correspond to curved vortex lines, provided they are bend so smoothly that any section of length $l$ is practically straight. These curved vortex lines, which represent most of the secondary maxima in Bell's functional integral, will either start and end on the walls of the volume $\Omega$, or they will close in on themselves to form vortex rings. In the thermodynamic limit $\Omega \to \infty$ one can obviously disregard those vortex lines which go from wall to wall; hence only the vortex rings are important.

The next limitation on the set of secondary maxima in Bell's integral which will be taken into account to calculate the partition function, is that we shall consider vortex rings which carry only *one* quantum of circulation $h/m$. This approximation suggests itself if one realizes that many more ways exist to divide an amount $nh/m$ of circulation over $n$ rings, each of which carries one quantum $h/m$, than over one ring which carries $n$ quanta. This is so essentially because $n$ lines can be drawn inside $\Omega$ in far more ways than one line. This tendency of the system to distribute the circulation over many vortex lines carrying one quantum each has also been observed in the experiments on rotating helium reported in [10, 12, 13].

Finally one should keep in mind that any two elements of vortex rings should have a spatial separation of the order of the healing length because it is only in this case that the flow pattern can be a solution of the stationarity equation.

After the preceeding remarks we now evaluate the partition function in a way similar to the method of Section II.4 (cf. Eq. II.4.1). The various steps are as follows. First, one notes that the contribution of the absolute maximum to the partition function is found upon substitution of (2.2) into (2.1):

$$\exp\{-L[\hat{\Psi}_0]\} = \exp\left\{\frac{\beta\mu^2}{2\bar{V}(0)}\,\Omega\right\} . \tag{4.9}$$

Second, one calculates the contribution of a piece of vortex line of length $Nl$ to the partition function. This can be done by substituting the dimensionless variables (1.12) and (1.13) with $n = 1$ in (4.6):

$$\rho \equiv r/l \ , \quad h \equiv (\bar{V}(0)/\mu)^{1/2} f \ , \tag{4.10}$$

which leads to the equation

$$\frac{d^2h}{d\rho^2} + \frac{1}{\rho}\frac{dh}{d\rho} - \frac{h}{\rho^2} + h - h^3 = 0 \ , \tag{4.11}$$

with boundary conditions

$$h(0) = 0 \ , \quad h(\infty) = 1 \ . \tag{4.12}$$

This function $h(p)$ should not be confused with Planck's constant. Because of the boundary condition at infinity one can find the asymptotic behavior of $h$ from the ansatz

$$h(\rho) \cong 1 + B\rho^{-\alpha} \ , \quad \alpha > 0 \ , \quad (\rho \gg 1) \ . \tag{4.13}$$

Substitution into (4.11) and comparison of the leading terms gives $\alpha = 2$, $B = -\frac{1}{2}$ so

$$h(\rho) \cong 1 - \frac{1}{2}\rho^{-2} \ , \quad (\rho \gg 1) \ . \tag{4.14}$$

Now let $\hat{\Psi}_{1,N}$ denote the solution of the stationarity equation with one vortex of length $Nl$. Its contribution to the partition function is found from (2.1):

$$\exp\{-L[\hat{\Psi}_{1,N}]\} = \exp\left\{\frac{\beta\mu^2}{2\bar{V}(0)}\Omega - \frac{\beta\bar{V}(0)}{2}\int_\Omega [A^4(\infty) - A^4(\mathbf{r})]\,d^3r\right\} . \tag{4.15}$$

Substitution of the new variables (4.10) and transformation of the space integral to cylindrical coordinates give

$$\exp\{-L[\hat{\Psi}_{1,N}]\} = \exp\left\{\frac{\beta\mu^2}{2\tilde{V}(0)}\Omega - Nl\frac{\pi\mu\beta\hbar^2}{2m\tilde{V}(0)}\int_0^{R/l}(1-h^4)\rho d\rho\right\} . (4.16)$$

Here the cut-off $R$ is of the order of the average distance between the vortex filament and the nearest part of another vortex. When the asymptotic behavior (4.14) is substituted one finds that the integral in (4.16) diverges weakly (logarithmically) if the upper limit $R/l \to \infty$. The leading asymptotic term is

$$\int_0^{R/l}(1-h^4)\rho d\rho \cong 2\ln\frac{R}{l}, \qquad (R/l \gg 1) . \qquad (4.17)$$

Combination of the last two equations gives for the contribution of a vortex filament of length $Nl$ the expression

$$\exp\{-L[\hat{\Psi}_{1,N}]\} = \exp\left\{\frac{\beta\mu^2}{2\tilde{V}(0)}\Omega\right\}\exp\{-c(\mu)\mu^{1/2}N\} , \qquad (4.18)$$

where

$$c(\mu) = \frac{\pi}{\sqrt{2}}\beta\hbar^3\,\tilde{V}(0)^{-1}\,m^{-3/2}\ln\frac{R}{l} \qquad (4.19)$$

is only weakly dependent on $\mu$. Note that the vortex weight (4.18) is of the non-analytic form

$$\text{(weight)} = \exp\{-(\text{interaction energy})^{-1}\} , \qquad (4.20)$$

which could never have been derived by means of perturbation theory. This once again shows the great power of the method of path integration.

The third step in our calculation of the partition function is to consider a solution of the stationarity equation which corresponds with a flow pattern in which an arbitrary number of closed vortex rings are present, with a total circumference $Nl$. Let $W(N)$ denote the number of different configurations of directed closed rings which have a total length $Nl$. The combined weight of these $W(N)$ different vortex configurations equals

$$\exp\left\{\frac{\beta\mu^2}{2\tilde{V}(0)}\Omega\right\}W(N)\exp\{-c(\mu)\mu^{1/2}N\} . \qquad (4.21)$$

Now, the partition function can be approximated by the sum of these weights, for a varying total length of all vortex filaments. This gives the expression

$$Z(\mu, \beta, \Omega) = \exp\left\{\frac{\beta\mu^2}{2\tilde{V}(0)}\,\Omega\right\} \sum_{N=0}^{\infty} W(N)\,\exp\{-c(\mu)\mu^{1/2}N\} \ , \quad (4.22)$$

the further evaluation of which is the subject of the rest of this chapter.

The detailed analysis depends crucially on our knowledge of the combinatorial factor $W(N)$ which, unfortunately, cannot be calculated rigorously. Before turning to an approximate evaluation it should be pointed out that, if $W(N)$ were known explicitly, the thermodynamic functions of the interacting Bose fluid could be found rigorously. This is the case because all terms in the sum (4.22) are positive and an upper limit to the number of terms is of the order $3\Omega/l^3$. Let the largest term occur for $N = N^*(\mu, \beta, \Omega)$, so

$$W(N^*)\,\exp(-c\mu^{1/2}N^*) = \max_{N} W(N)\,\exp(-c\mu^{1/2}N) \ . \quad (4.23)$$

The obvious inequality

$$W(N^*)\,\exp(-c\mu^{1/2}N^*) < \sum_{N=0}^{\infty} W(N)\,\exp(-c\mu^{1/2}N)$$

$$< \frac{3\Omega}{l^3}\,W(N^*)\,\exp(-c\mu^{1/2}N^*) \quad (4.24)$$

leads, after taking the logarithm and division by $\Omega$, to the grand canonical pressure

$$\beta p(\mu, \beta) = \frac{\beta\mu^2}{2\tilde{V}(0)} + \lim_{\Omega\to\infty}\frac{1}{\Omega}[\ln W(N^*) - c(\mu)\mu^{1/2}N^*] \ . \quad (4.25)$$

This shows (cf. (2.4) and (2.5)) that the vortex ring model gives, in principle, all the thermodynamic functions, provided $\mu$ is positive.

## 8.5   The independent vortex-ring model

It is now quite natural to suggest that the collective phenomenon which underlies the transition from the superfluid to the normal phase of the Bose liquid is due to an order-disorder transition in the system of vortex rings. In this section, which still follows Ref. [9], we shall use an approximation for $W(N)$ to explore the consequences of this idea.

The combinatorial problem is somewhat similar to the combinatorics of a large number of polymer rings, various aspects of which were considered in Chapters III and IV. This suggests approximations in which we relax some of the constraints on the vortex filaments. For example, the combinatorial problem would be completely trivial if there were no constraints on the vortex filaments at all. A better approximation would be to neglect the constraints between different vortex rings, but to keep the constraints on the shape of a single vortex ring (the independent vortex-ring model).

Hence, each vortex-ring can be represented by a closed random walk of some length. The constraint that any two elements of vortex rings should be separated by at least a distance $l$ translates into the constraint of the self-avoiding nature of these random walks. Let $u_n$ denote the number of self-avoiding closed random walks in a continuous space of total length $nl$, which start and end in some given point. In the independent vortex-ring model Eq. (4.22) now assumes the form

$$Z(\mu, \beta, \Omega) = \exp\left\{\frac{\beta\mu^2}{2\bar{V}(0)}\Omega\right\}$$

$$\times \left\{1 + \left(\frac{\Omega}{l^3}\sum_{n=1}^{\infty}\frac{u_n}{n}e^{-c\mu^{1/2}n}\right) + \frac{1}{2!}\left(\frac{\Omega}{l^3}\sum_{n=1}^{\infty}\frac{u_n}{n}e^{-c\mu^{1/2}n}\right)^2 + \ldots\right\}. \quad (5.1)$$

The combinatorial factors $1/k!$ and $1/n$ are due to the fact that the vortex-rings must be counted as geometrical objects, whereas the closed random walks have identities and their steps can be labelled individually. Using (2.4) one finds for the pressure

$$\beta p(\mu, \beta) = \frac{\beta\mu^2}{2\bar{V}(0)} + \frac{1}{l^3}\sum_{n=1}^{\infty}\frac{u_n}{n}\exp(-c\mu^{1/2}n) . \quad (5.2)$$

The successive terms between the braces on the right-hand side of (5.1) represent the relative weights of all configurations in which $0, 1, 2, \ldots$ vortex rings are present in the flow field. From these weights it is straightforward to calculate the average number $(\nu)$ of vortex rings in the volume. One finds

$$\nu(\mu, \beta, \Omega) = \frac{\Omega}{l^3}\sum_{n=1}^{\infty}\frac{u_n}{n}\exp(-c\mu^{1/2}n) . \quad (5.3)$$

Another physically interesting quantity is the total length $(\Lambda)$ of all (ring-shaped) vortex filaments. Equation (4.22) shows that this quantity equals

$$\Lambda(\mu, \beta, \Omega) = -l\frac{\partial \ln Z}{\partial(c\mu^{1/2})} . \quad (5.4)$$

But with (2.4) and (5.2) this gives

$$\Lambda(\mu, \beta, \Omega) = -l\Omega \, \frac{\partial \beta p}{\partial (c\mu^{1/2})}$$

$$= \frac{\Omega}{l^2} \sum_{n=1}^{\infty} u_n \exp(-c\mu^{1/2}n) \, . \tag{5.5}$$

We are now in a position to discuss the physical features of this model, as contained in the last four equations.

First, we note that both the total number and total length of all vortex-rings are extensive quantities. The intensive quantities $\nu/\Omega$ and $\Lambda/\Omega$ decrease monotonically with increasing $\mu$, once $\mu$ exceeds a certain value. This leads to the first qualitative conclusion: if we increase the density of the system at constant temperature (i.e., if we bring the system far into the superfluid phase) $\mu$ increases (cf. Eq. (2.5)) so the number of vortex rings will decrease to zero, as will their total length. This identifies the superfluid phase as a phase in which only very few, short vortex rings occur in the system; the fewer there are the farther away from the lambda transition the system is.

Now, suppose we keep the system in the superfluid phase but approach the transition to the normal phase by decreasing the density, i.e., by decreasing $\mu$. In this case, $\exp(-c\mu^{1/2})$ will increase to the value which corresponds to the radius of convergence $\mu_c$ of the infinite series in (5.2, 3, 5), the vortex rings will become more numerous and their total length will increase.

For $\mu < \mu_c$ these series diverge. We shall identify this region with the normal phase of the interacting Bose fluid. The divergence of the series in $p$, $\nu$ and $\Lambda$ for $\mu < \mu_c$ indicates that in the normal phase the flow field is unstable for the formation of vortex filaments; therefore the approximation of treating the different vortex rings as independent breaks down. We have seen that, if the constraints between different vortex rings could be taken into account, a finite expression (4.25) for the pressure would have arisen. Hence, only a qualitative conclusion holds: in the normal phase $\mu < \mu_c$ the flow field is laced densely with very long vortex lines.

A very rough and qualitative picture of the transition of the superfluid to the normal phase can be found as follows. The number $u_n$ of self-avoiding closed random walks of length $nl$ which start in some given point was discussed in Section 4.4 and given by (IV.4.1)

$$u_n \cong A\lambda^n n^{-a} \, , \qquad (n \gg 1) \, , \tag{5.6}$$

$$a \cong 7/4 \, . \tag{5.7}$$

Substitution into (5.2) gives for the pressure

$$\beta p(\mu, \beta) = \frac{\beta \mu^2}{2\bar{V}(0)} + \frac{A}{l^3} \sum_{n=1}^{\infty} n^{-a-1} e^{-n\xi} , \qquad (5.8)$$

$$\xi = c(\mu)\mu^{1/2} - \ln \lambda . \qquad (5.9)$$

A small error is made by using (5.6) also for $n$ of order unity, but this will not change the qualitative features of the theory, like the order of the phase transition. This order-disorder transition is somewhat hidden in Eq. (5.8) but shows up more clearly in the higher derivatives of the pressure with respect to $\mu$. Its location is at such a value $\mu_c$ of $\mu$ that $\xi=0$, or

$$c(\mu_c) \, \mu_c^{1/2} = \ln \lambda . \qquad (5.10)$$

When $\mu \downarrow \mu_c$, $\xi \downarrow 0$ and the series in (5.8) is still convergent. The same holds for the grand canonical density $\rho$ (cf. (2.5))

$$\beta \rho(\mu, \beta) = \frac{\partial \beta p(\mu, \beta)}{\partial \mu} = \frac{\beta \mu}{\bar{V}(0)} - \frac{A}{l^3} \frac{d\xi}{d\mu} \sum_{n=1}^{\infty} n^{-a} e^{-n\xi} , \qquad (5.11)$$

but the next derivative

$$\frac{\partial^2 \beta p(\mu, \beta)}{\partial \mu^2} = \frac{\beta}{\bar{V}(0)} - \frac{A}{l^3} \frac{d^2\xi}{d\mu^2} \sum_{n=1}^{\infty} n^{-a} e^{-n\xi}$$

$$+ \frac{A}{l^3} \left( \frac{d\xi}{d\mu} \right)^2 \sum_{n=1}^{\infty} n^{-a+1} e^{-n\xi} \qquad (5.12)$$

contains a term (the last one) which diverges to $+\infty$ when $\mu$ decreases to $\mu_c$. The second derivative of the pressure occurs, among other terms, in the specific heat. Hence the divergent part of $C_V(\mu, \beta)$ is proportional to (5.12) and we have to investigate the infinite series which appears here.

A full analysis of these series has been given by Robinson [26], Truesdell [27], Fisher [28] and Erdelyi *et al.* [29]. The essential idea is the following. For $0 < \xi \ll 1$ the successive values of $x \equiv n\xi$ are very close and one can approximate the sum by an integral. This gives the asymptotic formula

$$\sum_{n=1}^{\infty} n^{-a+1} e^{-n\xi} \cong \xi^{a-2} \int_0^{\infty} x^{-a+1} e^{-s} dx$$

$$(5.13)$$

$$= \xi^{a-2} \Gamma(2-a)$$

in terms of the gamma function (Ref. IV.9 Eq. 6.1.1). Integrating once from 1 to $\xi$, one finds

$$\sum_{n=1}^{\infty} n^{-a} e^{-n\xi} \cong \sum_{n=1}^{\infty} n^{-a} - \frac{\Gamma(2-a)\, \xi^{a-1}}{a-1} \ . \tag{5.14}$$

In a similar way the reader can find an asymptotic expansion for the series in (5.8)

When the last four equations are combined one finds that the divergent part of the specific heat $C_V$ is asymptotically proportional to

$$C_V(\mu, \beta) \sim (\mu - \mu_c)^{a-2} \ , \qquad (\mu \downarrow \mu_c) \ . \tag{5.15}$$

In the same way the density is seen to be asymptotically proportional to

$$\rho - \rho_c(T) \sim (\mu - \mu_c)^{a-1} \ , \qquad (\mu \downarrow \mu_c) \ , \tag{5.16}$$

where $\rho_c(T)$ is the density in the transition point. Elimination of $\mu - \mu_c$ between the last two equations gives the singularity of the specific heat as a function of $\rho$ and $T$

$$C_V(\rho, T) \sim (\rho - \rho_c)^{(a-2)/(a-1)} \ , \ (\rho \downarrow \rho_c) \ . \tag{5.17}$$

For an experiment in which one keeps $\rho$ constant but varies $T$ this gives a divergence

$$C_V(T) \sim (T_\lambda - T)^{(a-2)/(a-1)} \ , \ (T \uparrow T_\lambda) \ , \tag{5.18}$$

where now $T_\lambda$ is that value of $T$ for which

$$\rho_c(T_\lambda) = \rho \ . \tag{5.19}$$

Traditionally one writes the divergence of a specific heat in the form

$$C_V(T) \sim (T_\lambda - T)^{-\alpha} \ , \qquad (T \uparrow T_\lambda) \ , \tag{5.20}$$

so we have found a relation between the specific heat critical exponent $\alpha$ and the self-avoiding random walk exponent $a$

$$\alpha = \frac{2 - a}{a - 1} \cong \frac{1}{3} \ . \tag{5.21}$$

This value 0.33 is somewhat larger than the experimental value for $\alpha$, which is closer to 0, corresponding to an almost logarithmic singularity in the specific heat. Of course, the independent vortex ring model is only a very crude approximation to the summation of the full series (4.22), and an improved evaluation of the combinatorial factor $W(N)$ is still highly desirable.

# References

[1] E.P. Gross, in *Physics of Many-Particle Systems*, E. Meeron, ed. (Gordon and Breach, New York, 1966) p. 231.
[2] F.W. Wiegel, Ref. I–13, p. 419.
[3] F.W. Wiegel and J.B. Jalickee, *Physica* **57** (1972) 317.
[4] A.E. Glassgold, A.N. Kaufman and K.M. Watson, *Phys. Rev.* **120** (1960) 660.
[5] T.M. Rice, *Phys. Rev.* **140** (1965) A1889.
[6] J.S. Langer and M.E. Fisher, *Phys. Rev. Lett.* **19** (1967) 560.
[7] J.S. Langer, *Phys. Rev.* **167** (1968) 183.
[8] J.S. Langer and J.D. Reppy in *Progress in Low-Temperature Physics*, Vol. 6, C.J. Gorter, ed. (North-Holland, Amsterdam, 1970) p. 1.
[9] F.W. Wiegel, *Physica* **65** (1973) 321.
[10] J. Wilks, *The Properties of Liquid and Solid Helium* (Clarendon Press, Oxford, 1967).
[11] L. Onsager, *Nuovo Cimento* **6** Suppl. 2 (1949) 279.
[12] A. Fetter, *Boulder Lectures in Theoretical Physics* **XI B** (1968) 349.
[13] E.L. Andronikashvili and Y.G. Mamaladze, *Rev. Mod. Phys.* **38** (1966) 567.
[14] E. Byckling, *Ann. Phys.* **32** (1965) 367.
[15] V.N. Popov, *JETP* **64** (1973) 674.
[16] M. Rasetti and T. Regge, *Physica* **A80** (1975) 217.
[17] T. Banks, R. Myerson and J.B. Kogut, *Nucl. Phys.* **B129** (1977) 493.
[18] J.B. Kogut, A.I.P. Conf. Proc. **48** (1978) 17.
[19] S. Samuel, *Phys. Rev.* **D18** (1978) 1916.
[20] S. Samuel, *Nucl. Phys.* **B154** (1979) 62.
[21] P. Hasenfratz, *Phys. Lett.* **B85** (1979) 338.
[22] P. Hasenfratz, *Phys. Lett.* **B85** (1979) 343.
[23] R. J. Creswick and H.L. Morrison, *Phys. Lett.* **A76** (1980) 267.
[24] R.J. Creswick, "On the Application of the Functional Integral to the Interacting Bose Fluid," Ph.D. Dissertation, Dept. of Physics, Univ. of California, Berkeley, 1980 (unpublished).
[25] M. Cohen and R.P. Feynman, *Phys. Rev.* **107** (1957) 13.
[26] J.E. Robinson, *Phys. Rev.* **83** (1951) 678.
[27] C. Truesdell, *Ann. Math.* **46** (1945) 144.
[28] M.E. Fisher, *J. Chem. Phys.* **45** (1966) 1469.
[29] A. Erdelyi, W. Magnus, F. Oberhettinger and F.G. Tricomi, *Higher Transcendental Functions*, Vol. I, Sec. 1.11. (McGraw-Hill, New York, 1953).

# IX.   RENORMALIZATION GROUP THEORY OF
# THE INTERACTING BOSE FLUID

In the previous chapter we used increasingly more sophisticated forms of the saddle-point method to develop an approximate theory for the Bose fluid, especially for the low-temperature side of the lambda transition. In this chapter a quite different approximation scheme will be discussed: the method of the renormalization group. To date the renormalization group approach is the most powerful method for solving many-body problems, especially those in which a hierarchy of scales plays an essential role. Although several monographs exist which are devoted to the general ideas basic to this method [1–4] the application to the Bose fluid is still not generally appreciated. We shall, therefore, study this system in the following sections, presenting all derivations in some detail, and thereby providing a pedestrian introduction to the renormalization group.

For us the great interest in this method derives from the fact that it is a major step towards the establishment of a systematic calculus for path integrals. Actually, the way in which we evaluated the first path integral in this monograph (the cell representation for the Wiener integral of free Brownian motion, in Section 1.3) amounted to a kind of "weeding out" of the various integrations in the path integral, which is the basic idea of the renormalization method. It is, therefore, of interest to formulate this technique in as many different ways as possible, and to find that point of view from which it appears as simple as possible.

At the time of writing there already exist a considerable number of papers devoted to the subject of this chapter [5–22]; the earliest successful ones are due to K.K. Singh. We shall not discuss each paper in detail, there being a great variety of methods and formalisms. Instead we shall focus on the explicit, although approximate, calculation of the infinitesimal generator of the renormalization group, following Refs. 5 and 16.

Apart from finding the values of the critical exponents of the lambda transition most of the authors of Refs. 5–22 were led to the somewhat surprising result that these exponents are identical to those of a quite different classical system. Thus it appears that the universality of the critical exponents (i.e., their independence of the details of the model, apart from dimension) transcends the distinction between classical physics and quantum physics, a conclusion first reached by Singh about ten years ago. The fact that such a corresponding classical system exists must be a consequence of the collective behavior of a large number of bosons occupying the same quantum state. It should also be emphasized that the corresponding classical system is *not* a system of classical particles, but a classical two-component field theory., Hence, the quantum-mechanical origins of the λ-transition show up in the nature of the corresponding classical system, but not in the numerical values of the critical exponents. By contrast, a system of interacting fermions could probably not show this type of collective behavior and one should therefore expect its critical properties (which to the author's knowledge have not yet been calculated) to be quite different.

## 9.1  Short and long wavelengths

In order to set the scene for an explicit calculation one uses Eqs. (VII.6.16) and (VIII.1.2). For a cube of volume $\Omega = L^3$ the eigenfunctions $U_n(\mathbf{r})$ are plane waves, so

$$\Psi(\mathbf{r}, \tau) = (\Omega\beta)^{-1/2} \sum_{\mathbf{k},l} a_{\mathbf{k},l} \exp(i\mathbf{k} \cdot \mathbf{r} - 2\pi l\tau i/\beta) , \qquad (1.1)$$

where the components of $\mathbf{k}$ equal $(2\pi/L)$ times an integer because of the periodic boundary conditions at the surface of the volume. We introduce a cut-off in $\mathbf{k}$-space by restricting the $\mathbf{k}$-values to a sphere

$$|\mathbf{k}| < k_0 = \frac{2\pi}{a} , \qquad (1.2)$$

where $a$ denotes the scattering length of the interaction potential $V(\mathbf{r})$. We assume the potential to be predominantly repulsive, so $a > 0$. For hard spheres $a$ equals the diameter of the spheres. For a three-dimensional problem with this cut-off the potential can be replaced by the pseudopotential

$$V(\mathbf{r}) \cong \frac{4\pi a\hbar^2}{m} \delta(\mathbf{r}) ; \qquad (1.3)$$

cf. the detailed discussion in the monograph of Huang [23]. With this choice for the interaction the action (VII.6.8b) becomes, after a partial integration,

$$L[\Psi] = + \int \left[ \Psi^* \frac{\partial \Psi}{\partial \tau} + \frac{\hbar^2}{2m} |\nabla\Psi|^2 - \mu|\Psi|^2 + \frac{2\pi a\hbar^2}{m} |\Psi|^4 \right] dx \qquad (1.4)$$

This specifies the calculation problem with which one is faced.

With the substitutions

$$\tau = \beta\theta , \qquad (1.5a)$$

$$\Psi(\mathbf{r}, \tau) = \left( \frac{m}{\beta\hbar^2} \right)^{1/2} \phi(\mathbf{r}, \theta) , \qquad (1.5b)$$

the action can be written in the form

$$L[\phi(\mathbf{r}, \theta)] = \int_\Omega d^D r \int_0^1 d\theta \left\{ \frac{1}{2} |\nabla\phi|^2 + b_0 \phi^* \frac{\partial\phi}{\partial\theta} + U(\phi) \right\} \qquad (1.6)$$

where $D$ is the dimension of space and where

$$U(\phi) = \frac{1}{2} r_0 |\phi|^2 + s_0 |\phi|^4 , \qquad (1.7)$$

$$b_0 = \frac{m}{\beta\hbar^2} , \qquad (1.8)$$

$$r_0 = -\frac{2\mu m}{\hbar^2} , \qquad (1.9)$$

$$s_0 = \frac{2\pi a m}{\beta\hbar^2} . \qquad (1.10)$$

This transformation has eliminated the temperature dependence of the upper integration limit and it has made the coefficient in front of the gradient term independent of all the variables in the theory. Upon substitution of (1.5) into (1.1) one finds that the complex random field $\phi$ has the plane-wave expansion

$$\phi(\mathbf{r}, \theta) = \Omega^{-1/2} \sum_{\mathbf{k}, l} c_{\mathbf{k}, l} \exp\{i(\mathbf{k} \cdot \mathbf{r} - 2\pi l\theta)\} , \qquad (1.11)$$

where the expansion coefficients are

$$c_{\mathbf{k}, l} \equiv \left(\frac{\hbar^2}{m}\right)^{1/2} a_{\mathbf{k}, l} = c'_{\mathbf{k}, l} + ic''_{\mathbf{k}, l} . \qquad (1.12)$$

Substituting this in Eq. (VII.6.16b) the functional integration over $\Psi$ turns into the operation

$$\int d[\phi(\mathbf{r}, \theta)] \leftrightarrow \prod_{\mathbf{k}, l} \left\{ \frac{m}{\pi\hbar^2 \, \nu_{\mathbf{k}, l}} \int_{-\infty}^{+\infty} dc'_{\mathbf{k}, l} \int_{-\infty}^{+\infty} dc''_{\mathbf{k}, l} \right\} . \qquad (1.13)$$

The dimensions of these variables are $[\theta] = 1$, $[\phi] = [\text{length}]^{1 - D/2}$, $[c_{\mathbf{k}, l}] = [\text{length}]$.

It will turn out to be convenient to write (VII.6.16a,b) as an average

$$Z(z, \beta, \Omega) = N \left\langle \exp\left\{ - \int_\Omega d^D r \int_0^1 d\theta \, U(\phi) \right\} \right\rangle \qquad (1.14)$$

over the complex weight functional

$$W[\phi] = \prod_{\mathbf{k}, l} w(c_{\mathbf{k}, l}) , \qquad (1.15)$$

$$w(c_{\mathbf{k}, l}) = \frac{1}{\pi} \left(\frac{1}{2}|\mathbf{k}|^2 - 2\pi l b_0 i\right) \exp\left\{ - \left(\frac{1}{2}|\mathbf{k}|^2 - 2\pi l b_0 i\right) |c_{\mathbf{k}, l}|^2 \right\} . \qquad (1.16)$$

The normalization constant equals

$$N = \prod_{\mathbf{k}, l} \frac{m}{\hbar^2 \, \nu_{\mathbf{k}, l}} \left(\frac{1}{2}|\mathbf{k}|^2 - 2\pi l b_0 i\right)^{-1} . \qquad (1.17)$$

Substituting the value of $\nu_{\mathbf{k}, l}$ from (VII.6.11) the right-hand side becomes

$$N = \prod_{\mathbf{k}, l} \frac{\beta \xi_{\mathbf{k}} - 2\pi l i}{\beta \varepsilon_{\mathbf{k}} - 2\pi l i} \qquad (1.18)$$

where the $\varepsilon_{\mathbf{k}}$ are the free-boson energy eigenvalues (VIII.3.5). This product is similar to the one in (VII.6.12) and can be evaluated with the same methods. One finds

$$N = \prod_{\mathbf{k}} \frac{e^{-\beta\mu/2}}{1 - e^{-\beta\varepsilon_k/2}} . \tag{1.19}$$

This will add a term to $\beta p(z, \beta)$ equal to

$$\lim_{\Omega \to \infty} \frac{1}{\Omega} \ln N = -\frac{1}{2} \beta\mu(2\pi)^{-D}V_D(k_0)$$

$$- (2\pi)^{-D}\int \ln(1 - e^{-\beta\varepsilon_k/2}) \, d^Dk , \tag{1.20}$$

where $V_D(k_0)$ denotes the volume of a $D$-dimensional sphere of radius $k_0$.

After all these preliminaries we now introduce the long and short wavelength part of the complex random field by writing

$$\phi(\mathbf{r}, \theta) = \phi_l(\mathbf{r}, \theta) + \phi_s(\mathbf{r}, \theta) , \tag{1.21}$$

$$\phi_l(\mathbf{r}, \theta) = \Omega^{-1/2} \sum_{k<k_0-dk_0} \sum_l c_{\mathbf{k},l} \exp\{i(\mathbf{k}\cdot\mathbf{r} - 2\pi l\theta)\} , \tag{1.22}$$

$$\phi_s(\mathbf{r}, \theta) = \Omega^{-1/2} \sum_{k_0-dk_0<k<k_0} \sum_l c_{\mathbf{k},l} \exp\{i(\mathbf{k}\cdot\mathbf{r} - 2\pi l\theta)\} . \tag{1.23}$$

Here $dk_0$ is a small positive quantity which will vanish in a later stage of the calculation. The random field $\phi_s(\mathbf{r}, \theta)$ is a sum of plane waves with $\mathbf{k}$ values restricted to a shell in which $k_0 - dk_0 < |\mathbf{k}| < k_0$. Consequently the wavelengths contained in $\phi_s$ are of order $a$ because of (1.2). We shall call $\phi_s$ the short-wavelength part of the random field $\phi$. The random field $\phi_l(\mathbf{r}, \theta)$ is a sum of plane waves with $\mathbf{k}$ values anywhere in a sphere of radius $k_0 - dk_0$. Hence all wavelengths occur in $\phi_l$, but the typical one is of order $2a$. We shall call $\phi_l$ the long-wavelength part of the random field $\phi$. In the following sections we shall often make the assumption that the important wavelength in $\phi_l$ is large compared to $a$. This is only a rough approximation, but it simplifies the mathematics considerably.

The average (1.14) over the random function $\phi$ factorizes into the product of two partial averages, first over $\phi_s$, next over $\phi_l$:

$$< > = \ll >_s >_l , \tag{1.24}$$

in which the two partial averages are defined by (1.15) with the appropriate cut-offs in $\mathbf{k}$ space. The basic idea of the renormalization group approach is to perform the average over $\phi_s$ first and to write the $\phi_l$ average in a way similar to the original $\phi$ average.

The decoupling of the $\phi_s$ and $\phi_l$ averaging processes is suggested by the fact that any function $\phi_s$ is orthogonal to any function $\phi_l$. Hence

$$\int d^D r \int d\theta \left( \frac{1}{2} |\nabla \phi|^2 + b_0 \phi^* \frac{\partial \phi}{\partial \theta} \right) = \int d^D r \int d\theta \left( \frac{1}{2} |\nabla \phi_1|^2 + b_0 \phi_1^* \frac{\partial \phi_1}{\partial \theta} \right)$$

$$+ \int d^D r \int d\theta \left( \frac{1}{2} |\nabla \phi_s|^2 + b_0 \phi_s^* \frac{\partial \phi_s}{\partial \theta} \right) , \qquad (1.25)$$

which means that the weight factors for $\phi_s$ and $\phi_1$ also decouple. Next we realise that, because of the infinitesimal thickness $dk_0$ of the spherical shell in **k** space to which the **k** values in $\phi_s$ are restricted, the typical function $\phi_s$ can be considered to be small compared to the typical function $\phi_1$. This suggests expanding $U(\phi)$ in a Taylor series around $\phi_1$. Writing

$$\phi_s = \phi_s' + i\phi_s'' \qquad (1.26)$$

$$\phi_1 = \phi_1' + i\phi_1'' \qquad (1.27)$$

one finds

$$U(\phi) = U(\phi_1) + F(\phi_s) , \qquad (1.28)$$

$$F(\phi_s) = \sum_{n=1}^{\infty} F_n(\phi_s) , \qquad (1.29)$$

where $F_n$ is the sum of all terms of $n$th order in $\phi_s'$ and $\phi_s''$. For example, in an obvious notation,

$$F_1(\phi_s) = \phi_s' \left( \frac{\partial U}{\partial \phi'} \right)_1 + \phi_s'' \left( \frac{\partial U}{\partial \phi''} \right)_1 , \qquad (1.30)$$

$$F_2(\phi_s) = \frac{\phi_s'^2}{2!} \left( \frac{\partial^2 U}{\partial \phi'^2} \right)_1 + \phi_s' \phi_s'' \left( \frac{\partial^2 U}{\partial \phi' \partial \phi''} \right)_1 + \frac{\phi_s''^2}{2!} \left( \frac{\partial^2 U}{\partial \phi''^2} \right)_1 . \qquad (1.31)$$

The average over the short wavelength part of the random field thus leads to an expression of the form

$$\left\langle \exp \left\{ - \int U(\phi) d^D r \, d\theta \right\} \right\rangle_s = \qquad (1.32)$$

$$\exp \left\{ - \int U(\phi_1) \, d^D r \, d\theta \right\} \left\langle \exp \left\{ - \int F(\phi_s) \, d^D r \, d\theta \right\} \right\rangle_s .$$

Before we discuss the evaluation of the right-hand side with the cumulant expansion we shall study the properties of the short wavelength part ("rapid fluctuations") $\phi_s$ for their own sake in the next section, returning to the cumulants in Section 9.3.

## 9.2   Properties of the rapid fluctuations

It was stated already in the previous section that the complex weight functional $W_s$ of the rapid fluctuations $\phi_s$ is given by

$$W_s[\phi_s] = \prod_k{}'' \prod_l w(c_{k,l}) , \qquad (2.1)$$

with $w(c_{k,l})$ given by (1.16). Here the double prime indicates the constraints $k_0 - dk_0 < k < k_0$. It is straightforward to show that

$$\langle \phi_s(x) \rangle_s = \langle \phi_s^*(x) \rangle_s = 0 \qquad (2.2)$$

where the pair of variables $(\mathbf{r}, \theta)$ is denoted by $x$.

The rapid fluctuations are complex Gaussian random functions with an asymmetric covariance, as discussed in Section 7.5. Of the three two-point correlation functions two are found to be

$$\langle \phi_s(x)\, \phi_s(x') \rangle_s = \langle \phi_s^*(x)\, \phi_s^*(x') \rangle_s = 0 . \qquad (2.3)$$

More interesting is the covariance (cf. Eq. (VII.5.1 a, b, c))

$$G(x|x') = \langle \phi_s(x)\, \phi_s^*(x') \rangle_s \qquad (2.4)$$

$$= \Omega^{-1} \sum_k{}'' \sum_l \left( \frac{1}{2} k^2 - 2\pi l b_0 i \right)^{-1} \exp\{i\mathbf{k}\cdot(\mathbf{r}-\mathbf{r}') - 2\pi l(\theta - \theta')i\} .$$

This expression could be evaluated analytically with the theory of Section 7.5. It is perhaps more interesting to perform the calculations from scratch.

In order to do this one considers the integral

$$F_1(S) \equiv \frac{1}{2\pi i} \oint_C \frac{e^{-\theta t}}{(1-e^{-t})\,(S-t)}\, dt , \qquad (2.5)$$

where $S > 0$ and $0 < \theta < 1$. The contour of integration $C$ in the complex $t$-plane consists of two infinite straight lines, parallel to the imaginary $t$-axis, one crossing the real $t$-axis at some $t < 0$ and the other at some $t$ such that $0 < t < S$. The contour is traversed in a counter-clockwise direction. The integrand has poles at $t_l = 2\pi l i$ on the imaginary $t$-axis and a pole at $t = S$. Applying Cauchy's theorem of complex function theory we find that

$$F_1(S) = \sum_l (S - 2\pi l i)^{-1}\, e^{-2\pi l \theta i} . \qquad (2.6)$$

On the other hand, the integral vanishes exponentially if the real part of $t$

tends to $\pm\infty$. Hence $C$ can be deformed into a small loop $C'$ around the pole at $t = S$, traversed clockwise. In this case Cauchy's theorem gives

$$F_1(S) = \frac{e^{-\theta S}}{1 - e^{-S}} \ . \tag{2.7}$$

For $-1 < \theta < 0$ and $S > 0$ one considers the complex integral

$$F_2(S) = \frac{1}{2\pi i} \oint_C \frac{e^{-\theta t - t}}{(1 - e^{-t})(S - t)} \, dt \tag{2.8}$$

around the same contour. As it has the same residues in $t_l$ this integral is also given by (2.6). Also for Re $t \to \pm\infty$ the integrand again vanishes and the pole at $t = S$ gives

$$F_2(S) = \frac{e^{-\theta S - S}}{1 - e^{-S}} \ . \tag{2.9}$$

Combining the last five equations one finds

$$\sum_{l=-\infty}^{+\infty} \frac{e^{-2\pi l \theta i}}{S - 2\pi l i} = \begin{cases} \dfrac{e^{-\theta S}}{1 - e^{-S}} \ , & (0 < \theta < 1) \ , \\[4mm] \dfrac{e^{-\theta S - S}}{1 - e^{-S}} \ , & (-1 < \theta < 0) \ . \end{cases} \tag{2.10}$$

With the use of this **auxiliary** formula the covariance (2.4) can now be calculated easily. One finds

$$G(x|x') = \frac{dk_0}{b_0(2\pi)^D} \frac{e^{-(\theta - \theta')k_0^2/2b_0}}{1 - e^{-k_0^2/2b_0}} \oint e^{i\mathbf{k}\cdot(\mathbf{r}-\mathbf{r}')} \, d^{D-1}S \ , \tag{2.11}$$

for $\theta > \theta'$. Here the integral extends over the surface of a $D$-dimensional sphere in $\mathbf{k}$ space with radius $k_0$. For $\theta < \theta'$ one finds

$$G(x|x') = \frac{dk_0}{b_0(2\pi)^D} \frac{e^{-(\theta - \theta')k_0^2/2b_0 - k_0^2/2b_0}}{1 - e^{-k_0^2/2b_0}}$$

$$\times \oint e^{i\mathbf{k}\cdot(\mathbf{r}-\mathbf{r}')} \, d^{D-1}S \ . \tag{2.12}$$

Of course, the surface integrals in $\mathbf{k}$-space could be worked out for the various dimensions. We shall not do this as all that is needed to develop the theory further is the fact that $G$ is a function of $x - x'$ only, which vanishes with a long oscillating tail when $|\mathbf{r} - \mathbf{r}'| \gg k_0^{-1} = a/2\pi$. Of course, $G$ does not depend on the volume $\Omega$.

Next, we note that the rapid fluctuations also have the decomposition property (VII.5.2) which now reads

$$\left\langle \prod_{i=1}^{m} \{\phi_s(x_i)\} \prod_{j=1}^{n} \{\phi_s^*(x_j')\} \right\rangle_s = 0 , \qquad (m \neq n) , \tag{2.13}$$

$$= \sum_{P} \prod_{i=1}^{m} \langle \phi_s(x_i) \, \phi_s^*(x_{Pi}') \rangle_s , \qquad (m = n) ,$$

and which will be needed in the study of the cumulant expansion.

Finally, in order to estimate the order of magnitude of various terms in the cumulant expansion one has to evaluate various integrals of the covariance. Integrating (2.4) over the permitted values of $\mathbf{r}'$ and $\theta'$ one finds immediately that the integral of the covariance vanishes

$$\int G(x|x') \, dx' = 0 . \tag{2.14}$$

For the integral of $G^2$ one finds

$$\int G^2(x|x') \, dx' = \Omega^{-1} \sum_{\mathbf{k}}'' \sum_{l} \left( \frac{1}{2} k_0^2 - 2\pi l b_0 i \right)^{-1} \left( \frac{1}{2} k_0^2 + 2\pi l b_0 i \right)^{-1} \tag{2.15}$$

Using the summation formula (IV.2.23) this gives

$$\int G^2(x|x') \, dx' = C_D \, dk_0 , \tag{2.16a}$$

$$C_D = \frac{S_D(k_0)}{b_0 k_0^2 (2\pi)^D} \coth\left( \frac{k_0^2}{4b_0} \right) , \tag{2.16b}$$

where $S_D(k_0)$ equals the surface area of a $D$-dimensional sphere of radius $k_0$.

The integral of a higher power of $G$ can be estimated as follows. From (2.4) one has

$$\int G^n(x|x') \, dx' = \Omega^{-n+1} \sum' \prod_{j=1}^{n} \left( \frac{1}{2} |\mathbf{k}_j|^2 - 2\pi l_j b_0 i \right)^{-1} , \tag{2.17}$$

where $\Sigma'$ indicates a summation over $\mathbf{k}_1, l_1, \mathbf{k}_2, l_2, \ldots, \mathbf{k}_n, l_n$ such that

$$\sum_{j=1}^{n} \mathbf{k}_j = 0 , \tag{2.18a}$$

$$\sum_{j=1}^{n} l_j = 0 \; , \tag{2.18b}$$

in addition to the usual cut-offs $k_0 - dk_0 < k_j < k_0$. These two relations determine $\mathbf{k}_n$ in terms of the $(n - 1)$ other $\mathbf{k}_j$, and $l_n$ in terms of the $(n - 1)$ other $l_j$. As each $\mathbf{k}$ summation leads to one factor $dk_0$ we find the order of magnitude estimate

$$\int G''(x|x') \, dx' = O(dk_0)^{n-1} \; . \tag{2.19}$$

These properties of the rapid fluctuations will now be used to study the cumulant expansion.

## 9.3   The cumulant expansion

We return to the problem how to evaluate the average in (1.32), which is of the general form $<e^f>$ where $f$ is some stochastic variable. Replacing $f$ by $\lambda f$ one can expand the exponential in powers of $\lambda$ and average each term of the series

$$<e^{\lambda f}> = 1 + \lambda <f> + \frac{1}{2} \lambda^2 <f^2> + \dots \tag{3.1}$$

The right-hand side can again be written in the form of an exponential series:

$$= \exp\{\lambda <f> + \frac{1}{2} \lambda^2 (<f^2> - <f>^2) + O(\lambda^3 f^3)\} \; . \tag{3.2}$$

Setting $\lambda = 1$ one finds the first two terms of the cumulant expansion

$$<e^f> = \exp\{<f> + \frac{1}{2} (<f^2> - <f>^2) + O(f^3)\} \; . \tag{3.3}$$

Substitution of the explicit form of the stochastic variable from (1.32), and termination of the cumulant expansion after the second order term gives

$$<\exp\{- \int F(\phi_s) \, dx\}>_s \cong \exp\{- \int <F(\phi_s)>_s \, dx \tag{3.4}$$

$$+ \frac{1}{2} \int dx \int dx' \, [<F(\phi_s(x)) F(\phi_s(x'))>_s - <F(\phi_s(x))>_s <F(\phi_s(x'))>_s]\} \; .$$

In Ref. 5 the full cumulant expansion is studied. It is shown there that the corrections to the critical exponents due to the third and higher order cumulants are of order $\varepsilon^2$, where $\varepsilon = 4 - D$. In the remainder of this chapter we shall neglect these correction terms because a theory based on (3.4) is more transparant, and the results are fairly accurate anyhow.

The first cumulant is calculated by substituting (1.30) and (1.31) and using (2.2)

$$<F(\phi_s)>_s = \frac{1}{2} <\phi_s'^2(x)>_s \left(\frac{\partial^2 U}{\partial \phi'^2}\right)_1 + <\phi_s'(x) \phi_s''(x)>_s$$

$$\times \left(\frac{\partial^2 U}{\partial \phi' \partial \phi''}\right)_1 + \frac{1}{2} <\phi_s''^2(x)>_s \left(\frac{\partial^2 U}{\partial \phi''^2}\right)_1 + \ldots \quad (3.5)$$

The averages which appear here can be found by writing

$$\phi_s'(x) = \frac{1}{2} \{\phi_s(x) + \phi_s^*(x)\} \ ,$$

$$\phi_s''(x) = \frac{1}{2i} \{\phi_s(x) - \phi_s^*(x)\} \ , \quad (3.6)$$

and using (2.3) and (2.4). This gives

$$<\phi_s'(x) \phi_s''(x)>_s = 0 \ , \quad (3.7a)$$

$$<\phi_s'(x)^2>_s = <\phi_s''(x)^2>_s = \frac{1}{4} C_D k_0^2 dk_0 \ . \quad (3.7b)$$

The higher order terms, indicated by dots in (3.5), contain averages of four or more factors $\phi_s'$ or $\phi_s''$. Because of the decomposition property the averages are at least of order $(dk_0)^2$. Hence the first cumulant equals

$$-\int <F(\phi_s)>_s dx = -\frac{1}{8} C_D k_0^2 dk_0 \int \left(\frac{\partial^2 U}{\partial \phi'^2} + \frac{\partial^2 U}{\partial \phi''^2}\right)_1 dx + O(dk_0)^2 \ . \quad (3.8)$$

Since $U$ as given by (1.7) is not a function of $\phi'$ and $\phi''$ separately, but depends only on the modulus $|\phi| = (\phi'^2 + \phi''^2)^{1/2}$ this can also be written in the form

$$- \int <F(\phi_s)>_s \, dx = -\frac{1}{8} C_D k_0^2 \, dk_0$$

$$\times \int \left( \frac{d^2 U}{d |\phi|^2} + \frac{1}{|\phi|} \frac{dU}{d |\phi|} \right) dx + O(dk_0)^2 \quad (3.9)$$

No approximations had to be made in the derivation of this expression.

The situation is more complicated in the case of the second cumulant,

$$\frac{1}{2} \int dx \int dx' [<F(\phi_s(x))F(\phi_s(x')))>_s - <F(\phi_s(x))>_s <F(\phi_s(x')))>_s] \ . \quad (3.10)$$

This expression will be evaluated using the assumption that the scale on which $\phi_l$ varies appreciably in $\mathbf{r}$ space is very large as compared to the scale on which $\phi_s$ varies. In practice this means that we shall simplify integrals according to the recipe

$$\int G^n(x|x')f(\phi_l(x'))dx' \to f(\phi_l(x)) \int G^n(x|x') \, dx' \ , \quad (3.11)$$

for those functions $f(\phi_l)$ which will be encountered. As the scale of $G(x|x')$ is of the order $k_0^{-1} = a/2\pi$, according to (2.12), this implies that we assume the "typical" function $\phi_l(x)$ to remain practically constant within a region of linear dimension $a/2\pi$. This approximation is obviously somewhat crude, but it has the advantage that the resulting formalism remains rather simple. For a systematic improvement the reader might want to compare with Ref. 3.

As Eq. (3.9) shows that $<F(\phi_s)>_s$ is itself already of order $dk_0$ the second term in the integrand of (3.10) is $O(dk_0)^2$ and can be neglected as we are going to take the limit $dk_0 \to 0$ anyhow. Substituting the expansion (1.29) the second cumulant (3.10) is given by the double series

$$\frac{1}{2} \sum_{n,m=1}^{\infty} \int dx \int dx' <F_n(\phi_s(x))F_m(\phi_s(x')))>_s + O(dk_0)^2 \ . \quad (3.12)$$

In order to calculate the terms to order $dk_0$ one proceeds as follows:
(a)   $n = m = 1$: Substituting from (1.30), the term equals

$$<F_1(\phi_s(x))F_1(\phi_s(x'))>_s = <\phi_s'(x)\phi_s'(x')>_s \left( \frac{\partial U}{\partial \phi'} \right)_x \left( \frac{\partial U}{\partial \phi'} \right)_{x'}$$

$$+ <\phi_s''(x)\phi_s''(x')>_s \left( \frac{\partial U}{\partial \phi''} \right)_x \left( \frac{\partial U}{\partial \phi''} \right)_{x'} + <\phi_s'(x)\phi_s''(x')>_s \left( \frac{\partial U}{\partial \phi'} \right)_x \left( \frac{\partial U}{\partial \phi''} \right)_{x'}$$

$$+ <\phi_s''(x)\phi_s'(x')>_s \left( \frac{\partial U}{\partial \phi''} \right)_x \left( \frac{\partial U}{\partial \phi'} \right)_{x'} \quad (3.13)$$

in an obvious notation. Using (3.6, 7) and (2.14) it is seen that the integral over $x$ and $x'$ vanishes:

$$\frac{1}{2} \int dx \int dx' \, <F_1(\phi_s(x))F_1(\phi_s(x'))>_s = 0 \; . \tag{3.14}$$

(b)  $n=1$, $m=2$ and $n=2$, $m=1$: These terms lead to averages of products of three factors $\phi_s$. Hence they vanish rigorously by the symmetry of the weight functional. For the same reason all terms vanish for which $n+m$ is an odd number.

(c)  $n=m=2$: The calculation is straightforward in principle, but very tedious. One can, for example, rewrite the expression (1.31) for $F_2(\phi_s)$ in terms of $\phi_s$ and $\phi_s^*$ by considering $U$ as a function $U(\phi, \phi^*)$ of two independent variables $\phi$ and $\phi^*$:

$$F_2(\phi_s(x)) = \alpha(x)\phi_s^2(x) + \beta(x)\phi_s(x)\phi_s^*(x) + \gamma(x)\phi_s^{*2}(x) \; , \tag{3.15}$$

$$\alpha(x) = \frac{1}{2}\left(\frac{\partial^2 U}{\partial \phi^2}\right)_{\phi=\phi_1(x)} \; , \tag{3.16}$$

$$\beta(x) = \left(\frac{\partial^2 U}{\partial \phi \partial \phi^*}\right)_{\phi=\phi_1(x),\phi^*=\phi_1^*(x)} \; , \tag{3.17}$$

$$\gamma(x) = \frac{1}{2}\left(\frac{\partial^2 U}{\partial \phi^{*2}}\right)_{\phi^*=\phi_1^*(x)} \; . \tag{3.18}$$

Using a similar expression for $F_2(\phi_s(x'))$ one finds for the average of their product the sum of nine terms:

$$<F_2(\phi_s(x)) \, F_2(\phi_s(x'))>_s = \alpha\alpha'<\phi_s^2(x)\phi_s^2(x')>_s$$

$$+ \; \beta\beta'<\phi_s(x)\phi_s^*(x)\phi_s(x')\phi_s^*(x')>_s + \gamma\gamma'<\phi_s^{*2}(x)\phi_s^{*2}(x')>_s$$

$$+ \; \alpha\gamma'< \phi_s^2(x)\phi_s^{*2}(x')>_s + \gamma\alpha'<\phi_s^{*2}(x)\phi_s^2(x')>_s$$

$$+ \; \text{four other terms.} \tag{3.19}$$

Here $\alpha$, $\beta$, $\gamma$ have arguments $x$ and $\alpha'$, $\beta'$, $\gamma'$ have arguments $x'$.

The four other terms are averages of products of three factors $\phi_s$ and one factor $\phi_s^*$ or of one factor $\phi_s$ and three factors $\phi_s^*$. Because of the decomposition property (2.13) each of these four terms vanishes. For the same reason the first and third term on the right-hand side are equal to zero. The second term equals

$$\beta\beta' <\phi_s(x)\phi_s^*(x)\phi_s(x')\phi_s^*(x')>_s = \beta\beta'\{G(x|x)G(x'|x')$$
$$+ G(x|x')G(x'|x)\} , \quad (3.20)$$

the fourth term equals

$$\alpha\gamma'<\phi_s^2(x)\phi_s^{*2}(x')>_s = 2\alpha\gamma' \, G^2(x|x') , \quad (3.21)$$

and the fifth term equals

$$\gamma\alpha'<\phi_s^{*2}(x)\phi_s^2(x')>_s = 2\alpha'\gamma G^2(x'|x) . \quad (3.22)$$

Combination of the last four equations, and integration over $x'$ gives

$$\int <F_2(\phi_s(x))F_2(\phi_s(x'))>_s dx' = \int \left\{ \beta\beta' G(x|x')G(x'|x) \right.$$

$$\left. + 2\alpha\gamma' G^2(x|x') + 2\alpha'\gamma G^2(x'|x) + O(dk_0)^2 \right\} dx' , \quad (3.23)$$

where the $O(dk_0)^2$ term represents the first term on the right-hand side of (3.20) because $G(x|x) = O(dk_0)$.

At this point in the calculation one uses the approximation (3.11) to simplify the right-hand side of the last equation:

$$\int <F_2(\phi_s(x))F_2(\phi_s(x'))>_s dx' \cong \beta^2(x)\int G(x|x')G(x'|x) \, dx' +$$

$$2\alpha(x)\gamma(x) \int \{G^2(x|x') + G^2(x'|x)\} \, dx' + O(dk_0)^2 . \quad (3.24)$$

We shall simplify this expression even further by writing

$$\int G(x|x')G(x'|x) \, dx' \cong \int G^2(x|x') \, dx' . \quad (3.25)$$

A glance at (2.11, 12) shows that this holds for the $\mathbf{r}$, $\mathbf{r}'$ dependent part of $G$, but not for the $\theta$, $\theta'$ dependent part. However, it will be shown shortly that close to the lambda temperature the effective value of $b_0$ tends to infinity, which makes the function $G$ independent of $\theta$ and $\theta'$. Collecting these results, and using (2.16) for the integral of $G^2$, one finds

$$\int <F_2(\phi_s(x))F_2(\phi_s(x'))>_s dx' = \{\beta^2(x) + 4\alpha(x)\gamma(x)\} \, C_D \, dk_0 . \quad (3.26)$$

When (3.16–18) are substituted and when it is realized again that $U$ is not a function of $\phi$ and $\phi^*$ separately, but only of the modulus $|\phi|$, one finds for the term in (3.12) with $n = m = 2$

$$\frac{1}{2} \int dx \int dx' \, <F_2(\phi_s(x)) \, F_2(\phi_s(x'))>_s \cong$$

$$\frac{1}{16} \, C_D \, dk_0 \int \left[ \left( \frac{d^2 U}{d|\phi|^2} \right)^2 + \frac{1}{|\phi|^2} \left( \frac{dU}{d|\phi|} \right)^2 \right]_1 dx + O(dk_0)^2 \, . \quad (3.27)$$

(d)  $n = 1$, $m = 3$, $n = 3$, $m = 1$: These two terms lead to contributions proportional to $\int G(x|x') \, dx'$ which vanishes because of (2.14).

(e)  $n + m \geq 6$: These terms lead to contributions of the form

$$\int \int G^{C_1}(x|x) \, G^{C_2}(x|x') \, G^{C_3}(x'|x') \, dx \, dx' \; ; \quad 2(C_1 + C_2 + C_3) \geq 6 \, . \quad (3.28)$$

Because of (2.11, 12) and (2.19) this is of order $(dk_0)^{C_1+C_2+C_3-1}$; hence these terms are at least of second order in $dk_0$. These considerations show that the second cumulant (3.10) is given by the right-hand side of (3.27).

Substitution of the results (3.9) and (3.27) into (3.4) gives the explicit form of the cumulant expansion

$$<\exp\{-\int F(\phi_s) \, dx\}>_s \cong$$

$$\exp\left\{ -\frac{1}{8} \, C_D \, dk_0 \int \left[ k_0^2 \left( \frac{d^2 U}{d|\phi|^2} + \frac{1}{|\phi|} \frac{dU}{d|\phi|} \right)_1 \right. \right.$$

$$\left. \left. -\frac{1}{2} \left( \frac{d^2 U}{d|\phi|^2} \right)_1^2 - \frac{1}{2} \left( \frac{1}{|\phi|} \frac{dU}{d|\phi|} \right)_1^2 \right] dx + O(dk_0)^2 \right\} \, . \quad (3.29)$$

This somewhat unelegant expression answers the question we posed at the beginning of this section. It should be kept in mind that it could only be obtained making the two approximations (3.11) and (3.25), both of which are fairly accurate right at the lambda transition, but fail outside the critical region.

## 9.4  Scaling and the equivalence with a classical spin system

When the last equation of the previous section is combined with (1.14, 32) one finds that at this stage of the calculation the partition function is given by

$$Z(z, \beta, \Omega) = N <\exp\left\{ -\int U_1(\phi_1) \, dx \right\}>_1 \, , \quad (4.1)$$

where the average is taken with respect to the weight functional

$$W_1[\phi_1] = \frac{\exp\left\{-\int\left[\frac{1}{2}\mid\nabla\phi_1\mid^2 + b_0\ \phi_1^*\ \frac{\partial\phi_1}{\partial\theta}\right]dx\right\}}{\int\exp\left\{-\int\left[\frac{1}{2}\mid\nabla\phi_1\mid^2 + b_0\ \phi_1^*\ \frac{\partial\phi_1}{\partial\theta}\right]dx\right\}d[\phi_1(x)]}, \quad (4.2)$$

and where

$$U_1(\phi_1) = U(\phi_1) + \frac{1}{8}\ C_D\dot{}dk_0\left[k_0^2\left(U'' + \frac{U'}{\mid\phi_1\mid}\right)\right.$$

$$\left. -\frac{1}{2}\left(U''^2 + \frac{U'^2}{\mid\phi_1\mid^2}\right)\right] + O(dk_0)^2 . \quad (4.3)$$

Here the arguments on the right are $\mid\phi_1\mid$ throughout, and the prime denotes differentiation with respect to $\mid\phi_1\mid$. Essentially, the average over the short wavelength part $\phi_s$ of the random field has only led to an infinitesimal change in the function $U$ and to a change in the cut-off in **k** space.

The similarity between these formulae and the original expressions (1.14–16) can be improved by introducing "scaled" variables $\tilde{\mathbf{k}}, \tilde{\mathbf{r}}, \tilde{\phi}$ which are defined by

$$\tilde{\mathbf{k}} = \left(1 - \frac{dk_0}{k_0}\right)^{-1}\mathbf{k} , \quad (4.4a)$$

$$\tilde{\mathbf{r}} = \left(1 - \frac{dk_0}{k_0}\right)\mathbf{r} , \quad (4.4b)$$

$$\tilde{\phi}(\tilde{x}) = \left(1 - \frac{dk_0}{k_0}\right)^{1-D/2}\phi_1(x) . \quad (4.4c)$$

When (4.1–3) are transformed to the new variables they read

$$Z(z, \beta, \Omega) = \tilde{N} <\exp\left\{-\int\ \tilde{U}(\tilde{\phi})\ d\tilde{x}\right\}>_{\tilde{\phi}} . \quad (4.5)$$

The average is now with respect to the weight functional

$$\tilde{W}[\tilde{\phi}] = \frac{\exp\left\{-\int\left[\frac{1}{2}\mid\nabla\tilde{\phi}\mid^2 + \tilde{b}\ \tilde{\phi}^*\ \frac{\partial\tilde{\phi}}{\partial\theta}\right]d\tilde{x}\right\}}{\int\exp\left\{-\int\left[\frac{1}{2}\mid\nabla\tilde{\phi}\mid^2 + \tilde{b}\ \tilde{\phi}^*\ \frac{\partial\tilde{\phi}}{\partial\theta}\right]d\tilde{x}\right\}d[\tilde{\phi}(x)]} . \quad (4.6)$$

Here the gradients are with respect to $\tilde{r}$, the arguments are $\tilde{x}$ or $\tilde{\phi}$ and the values of $b_0$ and $\Omega$ are replaced by

$$\tilde{b} = \left(1 - \frac{dk_0}{k_0}\right)^{-2} b_0 , \qquad (4.7)$$

$$\tilde{\Omega} = \left(1 - \frac{dk_0}{k_0}\right)^{D} \Omega . \qquad (4.8)$$

There is no need to express $\tilde{N}$ in terms of $N$, which would be straightforward anyhow. The function $\tilde{U}$ is found to be

$$\tilde{U} = U + dk_0 \left[\frac{D}{k_0} U + \left(1 - \frac{D}{2}\right) \frac{|\tilde{\phi}|}{k_0} U' + \right.$$

$$\left. + \frac{1}{8} C_D k_0^2 \left(U'' + \frac{U'}{|\tilde{\phi}|}\right) - \frac{1}{16} C_D \left(U''^2 + \frac{U'^2}{|\tilde{\phi}|^2}\right)\right] + O(dk_0)^2 . \quad (4.9)$$

The primes denote differentiation with respect to $|\tilde{\phi}|$, and the arguments are $|\tilde{\phi}|$ everywhere. So, to summarize the calculation up till now, we have found that the effect of averaging over the rapid fluctuations and rescaling of the variables is threefold: (1) the volume decreases to $\tilde{\Omega}$; (2) the constant $b_0$ increases to $\tilde{b}$; (3) the "interaction" $U$ is "renormalized" into $\tilde{U}$. As both $\tilde{\phi}$ and $\tilde{x} \equiv (\tilde{r}, \theta)$ are integration variables we shall from here on omit the tilde on $\phi$ and $x$.

The equivalence of the critical behavior of the Bose fluid with that of a classical spin system now follows in a straightforward way. After a large number of renormalizations the effective value of the constant $b_0$ is transformed according to

$$b_0 \to b_1 \to b_2 \to \ldots \to b_n = \left(1 - \frac{dk_0}{k_0}\right)^{-2n} b_0 , \qquad (4.10)$$

and this eventually becomes very large compared to unity. The effective weight functional (4.6) can be written in the spectral form

$$\tilde{W} = \prod_{k<k_0} \prod_l w(c_{k,l}) , \qquad (4.11)$$

$$w(c_{k,l}) = \frac{1}{\pi} \left(\frac{1}{2} k^2 - 2\pi l \tilde{b} i\right) \exp\{-\left(\frac{1}{2} k^2 - 2\pi l \tilde{b} i\right) |c_{k,l}|^2\} . \qquad (4.12)$$

But as

$$\lim_{\tilde{b}\to\infty} w(c_{k,l}) = \begin{cases} \delta(c'_{k,l})\,\delta(c''_{k,l}) & \text{if } l \neq 0 , \quad (4.13) \\[2ex] \dfrac{1}{\pi}\left(\dfrac{1}{2}k^2\right)\exp\left\{-\dfrac{1}{2}k^2\,|c_{k,l}|^2\right\} & \text{if } l = 0 , \quad (4.14) \end{cases}$$

the only random fields which contribute in this case to the average (4.5) are those fields $\phi(\mathbf{r})$ which do not depend on $\theta$. In other words, close to the critical point, where the partition function is dominated by a very large number of renormalizations, it essentially has the functional integral representation

$$Z(z, \beta, \Omega) = (\text{constant}) \int \exp\left\{-\int\left[\frac{1}{2}|\nabla\phi|^2 + U(|\phi|)\right]d^D r\right\} d[\phi(\mathbf{r})] ,$$

$$(4.15)$$

where the integral is restricted to complex functions of $\mathbf{r}$ only. It follows from the theory developed in Chapter VI that such a functional integral describes a *classical* spin system in which the spin has two components. This demonstrates the famous equivalence between Bose- and classical critical behavior which we mentioned in the beginning of this chapter.

It should be kept in mind that: (a) the equivalence only holds true at the critical point; outside of the critical region the thermodynamic behavior of the interacting Bose fluid is of course entirely different from that of a classical system; (b) the ensembles are different, i.e., the Bose fluid in the grand canonical ensemble is related by (4.15) to the classical system in the canonical ensemble; (c) the equivalence is rigorous as it only depends on the fact that $b\to\infty$ after many renormalizations; the explicit result (4.9), which relies on various approximations, played no role in its derivation.

Before we end this section it should be noted for future reference that all the important quantities are well-behaved in the limit $b\to\infty$. For example, the quantity $C_D$ which occurs in (2.16) has the finite limiting value

$$\lim_{b\to\infty} C_D = \frac{4S_D(k_0)}{(2\pi)^D k_0^4} . \qquad (4.16)$$

Hence, after many renormalizations, $C_D$ can be treated as independent of the value of $b$.

## 9.5 Fixed points

The method of the renormalization group was developed by Wilson to study the critical behavior of systems consisting of interacting particles. In order to demonstrate this for interacting bosons, and to calculate the critical exponents, we first study the fixed points of the renormalization transformation

$$U(|\phi|) \to \tilde{U}(|\phi|) \tag{5.1}$$

specified by (4.9). This is an infinitesimal transformation which defines a "flow" in the space of "interactions" $U$. We shall always define $U$ and $\tilde{U}$ in such a way that

$$U(0) = 0 \; ; \qquad \tilde{U}(0) = 0 \; . \tag{5.2}$$

A glance at (4.5) and (4.9) shows that this can always be accomplished by absorbing the residual part of $\tilde{U}$ into the normalization constant $\tilde{N}$ by the replacement

$$N \to \tilde{N} \exp(-\Omega B d k_0) \; , \tag{5.3}$$

$$B = \lim_{\phi \to 0} \left[ \frac{1}{8} C_D \, k_0^2 \left( U'' + \frac{U'}{|\phi|} \right) - \frac{1}{16} C_D \left( U''^2 + \frac{U'^2}{|\phi|^2} \right) \right] \; . \tag{5.4}$$

The fixed points $U^*(|\phi|)$ of the renormalization transformation (5.1) are the solutions of the ordinary nonlinear differential equation

$$DU^* + \left( 1 - \frac{D}{2} \right) |\phi| \, U^{*\prime} + \frac{1}{8} C_D \, k_0^3 \left( U^{*\prime\prime} + \frac{U^{*\prime}}{|\phi|} \right)$$

$$- \frac{1}{16} C_D \, k_0 \left( U^{*\prime\prime 2} + \frac{U^{*\prime 2}}{|\phi|^2} \right) = K^* \; , \tag{5.5}$$

where $K^*$ is a constant which is determined by the differential equation itself, and which can have different values for different solutions $U^*$. The boundary conditions are

$$U^*(|\phi|) \to \infty \qquad \text{if} \qquad |\phi| \to \infty \; , \tag{5.6}$$

to guarantee the convergence of the average over $\phi$-space, and

$$U^{*\prime}(|\phi|) \to 0 \qquad \text{if} \qquad |\phi| \to 0 \; , \tag{5.7}$$

to guarantee that $U^*(|\phi|)$ stays finite if $|\phi| \to 0$.

The second boundary condition implies that $U^*$ is an even function of $|\phi|$. The first boundary condition tells us that $U^*$ should have the asymptotic behavior

$$U^*(|\phi|) \cong A \, |\phi|^g \; , \qquad (|\phi| \to \infty) \; , \tag{5.8}$$

where $A$ and $g$ have to be positive. Substituting into (5.5), and putting the sum of all terms of highest order equal to zero, one finds

$$g = 4 , \tag{5.9}$$

$$A = \frac{4-D}{10 C_D k_0} \quad \text{or } 0 . \tag{5.10}$$

As $A$ has to be positive one has found a very basic theorem: For $D > 4$ only the trivial fixed point solution $U^* = 0$ exists; for $D < 4$ at least one non-trivial fixed point exists.

This theorem shows that $D = 4$ is a special dimension in the sense that near $D = 4$ the nontrivial fixed point comes very close to the trivial one. This state of affairs has led Fisher and Wilson to what is called the $\varepsilon$-expansion: Treat $D$ as a continuous variable and put

$$D = 4 - \varepsilon , \quad (0 < \varepsilon \ll 1) . \tag{5.11}$$

Moreover, one writes the corresponding fixed point solution in the form

$$U^*(|\phi|) = \frac{1}{2} r^* |\phi|^2 + s^* |\phi|^4 + \dots , \tag{5.12}$$

where $r^* = O(\varepsilon)$, $s^* = O(\varepsilon)$, and the higher order terms have coefficients which are $O(\varepsilon^2)$ or smaller. Upon substitution into (5.5) one finds

$$r^* = -\frac{1}{5} k_0^2 \varepsilon + O(\varepsilon^2) , \tag{5.13}$$

$$s^* = +\frac{\varepsilon}{10 \, C_D \, k_0} + O(\varepsilon^2) , \tag{5.14}$$

$$K^* = -\frac{1}{20} C_D \, k_0^5 \varepsilon + O(\varepsilon^2) . \tag{5.15}$$

This non-trivial fixed point, which merges with the trivial one for $\varepsilon \downarrow 0$, is often called the Ising fixed point.

It will turn out to be of interest (for the calculation of the critical exponents in the next section) to study the renormalization transformation (5.1) in an infinitesimal vicinity of the Ising fixed point. This can be done by writing

$$U(|\phi|) = \frac{1}{2} r |\phi|^2 + s |\phi|^4 + \dots \tag{5.16}$$

with

$$r = r^* + \triangle r \ , \tag{5.17}$$

$$s = s^* + \triangle r \ , \tag{5.18}$$

where $\triangle r$ and $\triangle s$ are small. Substituting into (4.9) one finds that $\triangle r$ and $\triangle s$ are transformed into

$$\tilde{\triangle} r = \triangle r + 2 \left( \frac{\triangle r}{k_0} + 2C_D k_0^2 \triangle s - 2C_D r^* \triangle s - 2C_D s^* \triangle r \right) dk_0 \ , \tag{5.19}$$

$$\tilde{\triangle} s = \triangle s + \left( \frac{\varepsilon}{k_0} \triangle s - 20C_D s^* \triangle s \right) dk_0 \ , \tag{5.20}$$

where all terms of second or higher order in $\triangle r$ and $\triangle s$ have been neglected. In a more algebraic notation this can be written as

$$\begin{pmatrix} \tilde{\triangle} r \\ \tilde{\triangle} s \end{pmatrix} = R \begin{pmatrix} \triangle r \\ \triangle s \end{pmatrix} \tag{5.21}$$

where the explicit form of the 2x2 matrix $R$ is found from (5.13–14)

$$R = \begin{bmatrix} 1 + 2\left(1 - \dfrac{\varepsilon}{5}\right) \dfrac{dk_0}{k_0} & 4C_D\left(1 + \dfrac{\varepsilon}{5}\right) k_0^2 \, dk_0 \\[4mm] 0 & 1 - \varepsilon \dfrac{dk_0}{k_0} \end{bmatrix} \ . \tag{5.22}$$

The eigenvalues of this matrix are

$$\lambda_1 = 1 + 2\left(1 - \frac{\varepsilon}{5}\right) \frac{dk_0}{k_0} \ , \tag{5.23}$$

$$\lambda_2 = 1 - \varepsilon \frac{dk_0}{k_0} \ . \tag{5.24}$$

Hence, after $n$ renormalizations the effective interaction is transformed into

$$U_0 \rightarrow U_1 \rightarrow U_2 \rightarrow \ldots \rightarrow U_n \cong \frac{1}{2} r_n \, |\phi|^2 + s_n \, |\phi|^4 \ , \tag{5.25}$$

where

$$r_n \cong r^* + \lambda_1^n \triangle r_0 \ , \tag{5.26}$$

$$s_n \cong s^* + \lambda_2^n \triangle s_0 \ . \tag{5.27}$$

As $\lambda_1 > 1$ and $\lambda_2 < 1$ one thus finds, for $n \gg 1$,

$$r_n \cong r^* + \lambda_1^n \triangle r_0 , \tag{5.28}$$

$$s_n \cong s^* . \tag{5.29}$$

In the language of the renormalization group $r$ is a relevant variable (because $|r_n - r^*| \to \infty$ for $n \to \infty$ when $\triangle r_0 \neq 0$) and $s$ is an irrelevant variable (because $|s_n - s^*| \to 0$ even in the case $\triangle s_0 \neq 0$). Note that (4.10) shows that $b$ is also a relevant variable.

## 9.6 The critical exponents

The sequence (5.25) of effective "interactions" started with a function $U_0(|\phi(\mathbf{r})|)$ which itself results from the original expression (1.7) after a large number of renormalizations, i.e., sufficiently many to eliminate the $\theta$-dependence, as discussed in Section 9.4. Hence the $\triangle r_0$ and $\triangle s_0$ in (5.26, 27) are themselves still (rather complicated) functions of the variables $\mu$ and $T$ which occur in (1.6–10). Especially $\triangle r_0$ will be a function $f(\mu, T)$ of the chemical potential $\mu$ and the temperature $T$. Let $T_c(\mu)$ denote the temperature where, for a given fixed value of $\mu$, this function vanishes

$$f(\mu, T_c(\mu)) = 0 . \tag{6.1}$$

For $T$ near $T_c$ we have

$$\triangle r_0 \equiv f(\mu, T) = (T - T_c)\left(\frac{\partial f}{\partial T}\right)_{T=T_c} , \quad (|T - T_c| \ll T_c) . \tag{6.2}$$

In combination with (5.25, 28, 29) this shows that the effective interaction $U(|\phi|)$, which is of course also a function of $\mu$ and $T$ through the $\mu$, $T$ dependence of $r_n$ and $s_n$, has the property

$$U_n(\mu, T_c + \tau, |\phi|) = U_{n+1}\left(\mu, T_c + \frac{\tau}{\lambda_1}, |\phi|\right) , \tag{6.3}$$

provided $n \gg 1$. Now it should be remembered that after each renormalization the new unit of length equals the old unit of length times $(1 - dk_0/k_0)^{-1}$, as shown by Eq. (4.4b). The correlation length $\xi(\mu,T)$, which characterizes the spatial distance over which the fluctuations in the Bose fluid show significant correlations, is determined *as a function of the $n$-th unit of length* by the function $U_n(\mu, T, |\phi|)$. Hence (6.3) implies

$$\left(1 - \frac{dk_0}{k_0}\right)^n \xi(\mu, T_c + \tau) = \left(1 - \frac{dk_0}{k_0}\right)^{n+1} \xi\left(\mu, T_c + \frac{\tau}{\lambda_1}\right) , \tag{6.4}$$

simply because the left- (right-) hand side equals the correlation length in terms of the $n$-th ($n+1$-st) unit of length.

This functional relation for $\xi$ is sufficient to extract the physically relevant information. First, it shows that for $\tau = 0$ one has

$$\xi(\mu, T_c) = \left(1 - \frac{dk_0}{k_0}\right) \xi(\mu, T_c) \ . \tag{6.5}$$

This is a contradiction unless $\xi = 0$ or $\infty$, and as $\xi = 0$ is impossible we have $\xi(\mu, T_c) = \infty$. This identifies $T = T_c$ as the critical temperature, which is by definition the temperature at which the thermal fluctuations in the system show correlations over macroscopically large (=infinite!) distances. Second, the divergence of $\xi$ of $T \downarrow T_c$ is characterized by the critical exponent $\nu_0$ by means of

$$\xi(\mu, T) \sim (T - T_c)^{-\nu_0} \ , \qquad (\mu \text{ fixed}, T \downarrow T_c) \ . \tag{6.6}$$

Substitution into (6.4) gives

$$\tau^{-\nu_0} = \left(1 - \frac{dk_0}{k_0}\right) \left(\frac{\tau}{\lambda_1}\right)^{-\nu_0} \tag{6.7}$$

or, using Eq. (5.23),

$$\nu_0 = -\frac{\ln\left(1 - \dfrac{dk_0}{k_0}\right)}{\ln \lambda_1}$$

$$= \frac{1}{2} + \frac{\varepsilon}{10} + O(\varepsilon^2) \ . \tag{6.8}$$

This would give about $\nu_0 = 0.6$ for a three-dimensional system.

Now, the relation (6.6) also implies the divergence

$$\xi(\mu, T) \sim (\mu - \mu_c(T))^{-\nu_0} \ , \qquad (T \text{ fixed}, \mu \uparrow \mu_c) \ , \tag{6.9}$$

where $\mu_c(T)$ is the inverse of the function $T_c(\mu)$. The scaling hypothesis, on which the whole theory of critical phenomena relies, states that *the singular part of any thermodynamic function depends on $\mu$ and $T$ only through the combination $\xi(\mu, T)$*. Hence, the singular part of $\beta p(\mu, T)$, which has the same dimension as $\xi^{-D}$, must have the form

$$\beta p(\mu, T) \sim (\mu - \mu_c(T))^{+\nu_0 D} \ , \qquad (\mu \uparrow \mu_c) \ . \tag{6.10}$$

We verified already in Eqs. (VIII.5.11, 12) that this implies that the singular parts of density and specific heat must have the form

$$\rho(\mu, T) - \rho_c(T) \sim (\mu - \mu_c(T))^{\nu_0 D - 1} , \qquad (\mu \uparrow \mu_c) , \qquad (6.11)$$

$$C_V(\mu, T) \sim (\mu - \mu_c(T))^{\nu_0 D - 2} , \qquad (\mu \uparrow \mu_c) . \qquad (6.12)$$

When $\mu - \mu_c$ is expressed in terms of $\rho - \rho_c$ with the use of (6.11), and the result substituted into (6.9) and (6.12) one finds

$$\mu - \mu_c(T) \sim (\rho_c(T) - \rho)^{1/(\nu_0 D - 1)} , \qquad (6.13)$$

$$\xi(\rho, T) \sim (\rho_c(T) - \rho)^{-\nu_0/(\nu_0 D - 1)} , \qquad (6.14)$$

$$C_V(\rho, T) \sim (\rho_c(T) - \rho)^{(\nu_0 D - 2)/(\nu_0 D - 1)} , \qquad (6.15)$$

asymptotically for $T$ fixed, $\rho \uparrow \rho_c(\tau)$. Of course the same exponents will be found in experiments in which $\rho$ is kept fixed and $T$ varies. So, in the traditional notation

$$\xi(\rho, T) \sim (T - T_c)^{-\nu} , \qquad (\rho \text{ fixed}, T \downarrow T_c) , \qquad (6.16)$$

$$C_V(\rho, T) \sim (T - T_c)^{-\alpha} , \qquad (\rho \text{ fixed}, T \downarrow T_c) , \qquad (6.17)$$

we have found

$$\nu = \frac{\nu_0}{\nu_0 D - 1} = \frac{1}{2} + \frac{3\varepsilon}{20} + O(\varepsilon^2) , \qquad (6.18)$$

$$\alpha = \frac{2 - \nu_0 D}{\nu_0 D - 1} = \frac{\varepsilon}{10} + O(\varepsilon^2) . \qquad (6.19)$$

For $D = 3$ this gives $\nu \cong 0.65$ and $\alpha \cong 0.10$ by extrapolation. This is in fair agreement with the experiments on liquid $^4$He near the lambda point.

## References

[1] K.G. Wilson and J. Kogut, *Phys. Reports* **12** (1974) 75.
[2] G. Toulouse and P. Pfeuty, *Introduction au Groupe de Renormalisation et a ses Applications* (Presse Univ. de Grenoble, 1975). An English translation is also available.
[3] S.K. Ma, *Modern Theory* of *Critical Phenomena* (Benjamin, New York, 1976).
[4] C. Domb and M.S. Green, eds., *Phase Transitions and Critical Phenomena* (Academic Press, London, 1976), Vol. 6.
[5] R.J. Creswick and F.W. Wiegel, *Phys. Rev.* **A28** (1983) 1579.

[6]   S.K. Ma, *Phys. Rev. Lett.* **29** (1972) 1311.
[7]   F. Family and H.E. Stanley, *Phys. Lett.* **53A** (1975) 111.
[8]   K.K. Singh, *Phys. Lett.* **51A** (1975) 27.
[9]   K.K. Singh, *Phys. Rev.* **B12** (1975) 2819.
[10]  K.K. Singh, *Phys. Lett.* **57A** (1976) 309.
[11]  K.K. Singh, *Phys. Rev.* **B13** (1976) 3192.
[12]  A.D. Stella and F. Toigo, *Nuovo Cimento* **B34** (1976) 207.
[13]  M. Baldo, E. Catara and U. Lombardo, *Lett. Nuovo Cimento* **15** (1976) 214.
[14]  L.De Cesare, *Lett. Nuovo Cimento* **22** (1978) 325.
[15]  F.W. Wiegel, *Physica* **91A** (1978) 139.
[16]  F.W. Wiegel, Chapter I, Ref. 13, p.419.
[17]  J.C. Lee, *Phys. Rev.* **B20** (1979) 1277.
[18]  J.C. Lee, *Physica* **104A** (1980) 189.
[19]  G. Busiello and L. De Cesare, *Phys. Lett.* **77A** (1980) 177.
[20]  G. Busiello and L. De Cesare, *Nuovo Cimento* **B59** (1980) 327.
[21]  S.P. Ohanessian and A. Quattropani, *Helv. Phys. Acta* **46** (1973) 473.
[22]  S.P. Ohanessian, "The Interacting Bose Gas: Formalism in Coherent States and Applications," (Ph.D. Dissertation, Ecole Polytechnique Fédérale de Lausanne, 1976).
[23]  K. Huang, *Statistical Mechanics* (Wiley, New York, 1963).

# X.   PATH INTEGRALS AND THE HOLISTIC APPROACH TO THEORETICAL PHYSICS

The development of modern physics can be viewed from so many different directions that almost every author presents a different perspective. In this chapter we shall look at the very recent history of physics and try to discern the role which the concepts of path integration have played. Of course, we are interested in the future and in the way in which path integration might influence the future interpretation of reality. We follow Ref. 1.

To summarize the thesis of this chapter: it appears to us that theoretical physics is loosing its classical form of a local and deterministic theory and is acquiring a more global and organic form. Moreover, it will be argued that path integration is a very appropriate mathematical language for this new, more holistic form of physical theory. In order to clarify these matters we briefly sketch some relevant features of the recent history of physics.

## 10.1   Leibniz and the variational principles of mechanics

Simultaneously with the creation of physics as a modern science in the 17th century the mathematical tools were created with which physicists tried to describe reality. In fact people like Descartes and Newton, who were mathematicians as much as physicists, developed the branch of mathematics which even today is still known by the somewhat archaic name of the calculus of infinitesimals. Using this calculus one can formulate "laws" of nature which have a local and deterministic character. For example, Newton's equation of motion

$$m \frac{d^2\mathbf{r}}{dt^2} = \mathbf{F}(\mathbf{r}) \tag{1.1}$$

for a single point mass tells you that the position $\mathbf{r}$ and velocity $\mathbf{v}$ at time $t + \varepsilon$ can be calculated from the position and velocity at time $t$ by

$$\mathbf{r}(t + \varepsilon) \cong \mathbf{r}(t) + \varepsilon\mathbf{v}(t) \ , \tag{1.2}$$

$$\mathbf{v}(t + \varepsilon) \cong \mathbf{v}(t) + \frac{\varepsilon}{m} \mathbf{F}(\mathbf{r}(t)) \ . \tag{1.3}$$

By repeating this calculation many times over (by hand in the 17th century, with an electronic calculating device nowadays) the future trajectory of the particle can be predicted. This theory is deterministic in the sense that the whole future of the particle is uniquely determined by its position and velocity now. The theory is also local in the sense that its behavior at time $t$ is determined by a force field acting at the position of the particle at the same time $t$.

It is this theory, classical mechanics in its most elementary form, which still forms the basis of our understanding of the motions of planets and projectiles, gasses and fluids, and so on. If, for example, you are listening to a lecture, the sound waves caused by the speaker can be described by partial differential equations which summarize the changes in pressure and density of the air between times $t$ and $t + \varepsilon$. In the same way many natural phenomena in our daily life can be adequately explained with some appropriate generalization of Newton's law [2]. This might be the reason why the corresponding mechanistic concept of nature is still so well known among non-scientists although it has been obsolete in physics for almost a century.

Parallel with the development of the mechanistic theory of nature another development was initiated in the 17th century. This development, which eventually lead to the calculus of variations, is connected with the names of Leibniz, Euler and Lagrange. The motion of a point mass between times $t_1$ and $t_2$ is now not described by the differential equation (1.1) but by the variational principle

$$\delta \int_{t_1}^{t_2} L[\mathbf{r}(t), \dot{\mathbf{r}}(t)]\, dt = 0 \; , \tag{1.4}$$

where the Lagrangian

$$L \equiv \frac{1}{2}m\left(\frac{d\mathbf{r}}{dt}\right)^2 - V(\mathbf{r}(t)) \tag{1.5}$$

equals the difference between kinetic energy and potential energy. The variations of the integral are restricted to paths $\mathbf{r}(t)$ which are twice differentiable with respect to time and which pass through $\mathbf{r}_1$ at $t_1$ and through $\mathbf{r}_2$ at $t_2$

$$\delta\mathbf{r}(t_1) = \delta\mathbf{r}(t_2) = 0 \; . \tag{1.6}$$

From a purely mathematical point of view the variational principle (1.4–6) is strictly equivalent to Newton's law because the Euler-Lagrange equation for (1.4) is identical to (1.1). Yet, the new feature which the proponents of variational principles introduced into the description of nature is that (1.4) gives a global description of the motion of the particle; the whole trajectory of the particle through space-time must be known to decide if it obeys (1.4). In this sense the variational principles of classical mechanics are the first indication that a more global description of nature might be possible.

From where does this point of view arise? I cannot answer this question and am unaware of any special study of it; the best approximation to a historical study is Chapter IX of the monograph by Lanczos [3]. It seems striking that the first ideas relating to optimal principles in nature were formulated by G.W. Leibniz (1646–1716). What is striking about this is that Leibniz was the first European scientist who made a systematic study of the Chinese texts which were being translated into Latin by the members of the Jesuit mission in Peking.

The story of the Jesuits in China and their role in the two-way scientific exchange between China and Europe in the 17th century is an epic which has recently been put into a wide perspective by Needham and co-workers [4]. The first Jesuit to come to China was Matteo Ricci, who reached the imperial court in Peking in 1601. In the following years the members of the Jesuit mission translated many philosophical and scientific Chinese texts into Latin. At the European end of this scientific channel we find, among others, Athanasius Kircher, whose work Leibniz admired and expanded. This all suggests a historial development which starts with the Taoists in China and ends in modern theoretical physics, by way of Kircher and Leibniz. And most of this, 300 years ago!

Nowadays, three centuries later, it is well known how most of classical physics can be expressed in terms of variational principles [3, 5]. For a long time they were used as a mere calculational device. It is only with the development of the path-integral formulation of quantum physics that they have come back to a life of their own, expressing the essence of physical theory in the strange realm of Planck's constant.

## 10.2   Tetrode and the absorber theory of radiation

An analogous situation is found when one turns to another branch of theoretical physics: the theory of the electromagnetic field. The usual theory culminated around 1860 in Maxwell's equations [6]. These equations describe the electromagnetic field in vacuum (the radiation field) with the use of two vector fields, $\mathbf{E}$ and $\mathbf{H}$, the temporal development of which is determined by two first order equations (in Gaussian cgs units)

$$\text{curl } \mathbf{E} = -\frac{1}{c}\frac{\partial \mathbf{H}}{\partial t} , \qquad (2.1)$$

$$\text{curl } \mathbf{H} = +\frac{1}{c}\frac{\partial \mathbf{E}}{\partial t} + \frac{4\pi}{c}\mathbf{j} . \qquad (2.2)$$

Moreover, one has the two auxiliary equations

$$\text{div } \mathbf{E} = 4\pi\rho \qquad (2.3)$$

$$\text{div } \mathbf{H} = 0 . \qquad (2.4)$$

Here $\rho$ is the density of charge and $\mathbf{j}$ the density of current. Once again this theory is local and deterministic, and enables us to follow the development of the field in time by taking many small steps. Maxwell's equations lead to accurate predictions of electromagnetic phenomena and are still the conceptual basis of a large part of modern technology. They describe the lights in a lecture hall as well as the electromagnetic signals in the nervous system of an individual who is listening to a lecture.

In this case too, another development started around 1920 which eventually led to a totally different concept of electromagnetic radiation and to a global and almost organic description of nature. This development originated in the somewhat mysterious Dutch physicist Tetrode. In view of the historical character of this chapter and the extreme paucity of information about the life of Tetrode I summarize the little bit that is known to me, mainly through a recent article in a local newspaper [7].

Hugo Martin Tetrode was born in Amsterdam on March 7, 1895. At some unspecified date around 1910 he moved to Leipzig and studied mathematics, physics and chemistry. Tetrode must have been somewhat like a prodigy as his first paper appears in the *Annalen der Physik* of 1912. It is devoted to a problem in the molecular theory of heat and is a remarkable piece of work for a young man of 17. He returned to Amsterdam around 1914 and worked at home. All that is left from these years, during which elsewhere in Europe millions were massacring each other for no reason at all, are some short letters to Lorentz and Ehrenfest. After ten years (1922) a paper appeared which is now recognized as the first place in the literature where the ideas of the absorber theory of radiation are mentioned [8]. We shall comment upon these ideas shortly. Tetrode published two more papers in 1928, but tuberculosis, from which he must have suffered already before that time, worsened and eventually killed him on January 18, 1931. Whom the gods love dies young.

An anecdote about his lack of interest in personal exchanges with other physicists tells how once Einstein and Ehrenfest presented themselves at his house (this might have been Herengracht 526 in Amsterdam) but were turned back by the maid with the words, "Mr. Tetrode does not receive guests." I do not know if this ever happened, but it might explain why Einstein could still remember Tetrode's work on February 21, 1941, when the American Physical Society had a meeting in Cambridge (Massachusetts) where Wheeler and Feynman presented their ideas [9, 10]. They were at that time unaware of Tetrode's work but Einstein's comment led them to Ref. [8]. In a footnote to their first paper they also quote ideas of Ritz [11] and of Lewis [12]. There is no doubt that it is the work of Wheeler and Feynman which brought the absorber theory of radiation to full maturity.

In order to illustrate the difference between the theory of Maxwell and the absorber theory of radiation consider the following fictitious experiment. A charged particle is at rest in space (with respect to some inertial system of coordinates). At some time, for example at 4 pm one forces the particle to make a few oscillations, after which it is again left at rest in space. One now asks which electromagnetic signal is caused by these few oscillations, and first uses Maxwell's theory, next the absorber theory of radiation.

Using Maxwell's equations (2.1–4) with $\rho$ and $\mathbf{j}$ corresponding to the prescribed oscillations of the particle, one finds a solution which represents a spherically symmetric light wave which is centered on the particle and moves away from the particle with the speed of light. This exploding wave is usually called the retarded solution of the equations, as it exists at times later than 4 pm.

However, Maxwell's equations also have an advanced solution which describes an imploding spherical light wave, which is a spherical wave which travels through space at time earlier than 4 pm, towards the particle, and hits it at 4 pm. In the usual theory one omits the advanced solution as its existence would be in conflict with our ordinary concept of causality: if I shake an electric charge at 4 pm I expect an electromagnetic signal after 4 pm but not before that time!

Describing this same fictitious experiment with the absorber theory of radiation essentially amounts to accepting both the advanced and the retarded solutions as physical realities. According to the absorber theory the situation is as follows: at times before 4 pm a converging spherical electromagnetic wave travels through space towards the particle and hits it at 4 pm. As a result of the interaction of the charge with this field and some other external forces the particle makes a few oscillations and comes to rest shortly after 4 pm. At later times one finds another spherical light wave which diverges away from the particle.

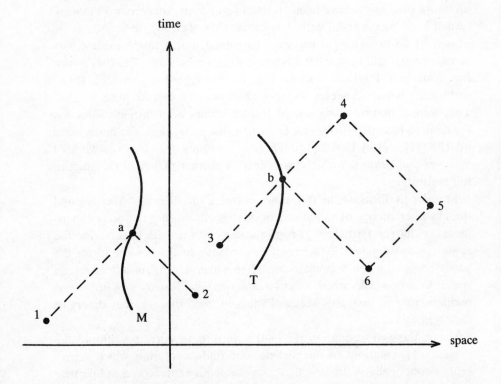

Fig. 10.1.  Space-time diagram indicating the causal connections in the Maxwell theory (obeyed by the particle M) and in Tetrode's theory (obeyed by particle T).

Obviously, the absorber theory leads to a picture of reality which differs considerably from the customary local and deterministic picture of Maxwell's theory. This is illustrated in Fig. 10.1 in which the space-time trajectories of two particles are drawn: M obeys the Maxwell theory, T obeys the absorber theory. In order to predict the trajectory of the first "Maxwell" particle at space-time point (a) one only has to know the locations (1,2) of other particles in its past light cone. However, to predict the trajectory of the second particle at space-time point (b) not only the location of other particles (3) in its past light cone are important, but also the location of other particles (4) in its future light cone. Actually, the "Tetrode" particle constantly interacts with itself too, by way of cycles of the type b, 4, 5, 6, b indicated in the figure. Here the movement of T in (b) influences the movement of another particle at (4) through its retarded field. This movement, through the advanced field of (4) rattles another particle at (5), the advanced field of which rattles (6), the retarded field of which will arrive back at (b) as a remote echo of its original movement.

It will be clear that it is very hard to do actual calculations with the absorber theory of radiation. It is almost a miracle that Wheeler and Feynman [9] went ahead and showed that in many situations the predictions of this theory are practically identical to those of Maxwell's theory! What is especially new and striking about the absorber theory is that it connects the richness of the electromagnetic phenomena of everyday life with the large-scale properties of the universe, and it is this aspect which has received most attention in recent years [13]. By way of illustration consider a visual observation of a remote galaxy, by means of a very large telescope, say a galaxy at a distance of about one billion light years. During such an observation charged particles in the retina of the observer's eye are influenced by the retarded fields which were emitted, a billion years ago, by certain charged particles in some of the stars of that galaxy. For reasons of time-symmetry the absorber theory now claims that the particles in the observer's retina are also subject to the advanced fields caused by accelerations of other charged particles a billion years in the future!

To summarize this section, the absorber theory of radiation leads to a picture of the universe in which the trajectories of all charged particles are coupled at all times, from the big bang to the big crunch, to each other and to themselves.

## 10.3 Path integrals and quantum physics

It will be clear from the discussion in Chapter V that the global aspect of quantum mechanics—the third main branch of theoretical physics—can be expressed by path integrals in an especially lucid fashion. The Schrödinger equation describes a particle with a wave function, the evolution of which can be calculated in small time steps. Instead, the Feynman path integral (V.2.4) describes the particle with an ensemble of paths $r(t)$ which fill space-time like some kind of very thin spaghetti. The many advantages of this point of view have formed the subject of this treatise. In this final section we only collect some further remarks.

(a) *Variational principles*: Even in the usual Schrödinger description the global aspect of quantum mechanics becomes obvious in the use of variational principles. The best known one is the statement that the ground-state energy of the system is given by

$$E_0 = \min \int \psi^* H\psi \, dx \; , \tag{3.1}$$

where the minimum is taken over all wave functions which are normalized to unity. Other quantum mechanical variational principles are discussed by Yourgrau and Mandelstam [5].

For real path integrals special variational principles exist; the most well known one is due to Feynman [I-34]. It is a consequence of the inequality

$$<e^x> \geq e^{<x>} \tag{3.2}$$

which holds for any real stochastic variable $x$ because of the concave nature of the graph of the function $e^x$. If now one has to evaluate any real path integral of the form

$$Z = \int \exp\{S[x(\tau)]\} \, d[x(\tau)] \; , \tag{3.3}$$

this is first written in the form of a product

$$Z = \int \exp\{S_0[x(\tau)]\} \, d[x(\tau)] \, \langle \exp(S - S_0) \rangle_{S_0} \; . \tag{3.4}$$

Here $S_0$ is some exponential such that its path integral can be evaluated analytically, and the second factor denotes an average with a weight functional $\exp\{S_0[x(\tau)]\}$. Combination of the last equation with (3.2) gives the inequality

$$Z \geq \int \exp\{S_0[x(\tau)]\} \, d[x(\tau)] \, \exp\{\langle S - S_0 \rangle_{S_0}\} \; . \tag{3.5}$$

In practical applications one uses a "trial functional" $S_0$ which still contains some adjustable constants $C_1$, $C_2$,... and maximizes the right-hand side by an appropriate choice of their values. This often leads to a very good numerical approximation of $Z$, even if the corresponding trial functional $S_0$ is a poor approximation of $S$. This method has been applied by Feynman and others with considerable success to the calculation of the mass of the polaron. The reader is refered to Ref. I–34, I–35, and I–13 for an extensive discussion of these applications.

(b) *The classical limit*: It is perhaps also remarkable from the point of view of holistic physics that the path-integral formulation of quantum mechanics enables one to go to the classical limit in such an elegant way, and that one recovers classical mechanics not in the Newtonian form (1.1) but in the variational form (1.4.5). The reader might want to pursue these matters in his own way, and—for the time being—the author will be silent.

## References

[1] F.W. Wiegel, Inaugural Lecture delivered at Twente University of Technology on 29 September 1983.

[2] M. Minnaert, *De Natuurkunde van 't Vrije Veld* (Thieme, Zutphen, 1974) Vols. I–III.

[3] C. Lanczos, *The Variational Principles* of *Mechanics* (University of Toronto Press, Toronto, 1949 and Oxford University Press, Oxford, 1952, 1957).

[4] J. Needham, *Science and Civilization in China* (Cambridge University Press, Cambridge, 1980) Vol. II, especially Section 16f.

[5] W. Yourgrau and S. Mandelstam, *Variational Principles in Dynamics and Quantum Theory* (Dover, New York, 1979).

[6] J.C. Maxwell, *Trans. Camb. Phil. Soc.* **10** (1856).

[7] H.B.G. Casimir, NRC-Handelsblad, February 23, 1984 (in Dutch).

[8] H. Tetrode, *Z. Phys.* **10** (1922) 317.

[9] J.A. Wheeler and R.P. Feynman, *Rev. Mod. Phys.* **17** (1945) 157.

[10] J.A. Wheeler and R.P. Feynman, *Rev. Mod. Phys.* **21** (1949) 425.

[11] W. Ritz, *Ann. d. Chem. et d. Physique* **13** (1908) 145.

[12] G.N. Lewis, *Proc. Nat. Acad. Sci.* **12** (1926) 22.

[13] F. Hoyle and J.V. Narlikar, *Action at a Distance in Physics and Cosmology* (Freeman and Company, San Francisco, 1974).